Elements in the Philosophy of Science
edited by
Jacob Stegenga
*NTU Singapore*

# THE PHILOSOPHY OF LINGUISTICS

Ryan M. Nefdt
*University of Cape Town and University of Bristol*

Shaftesbury Road, Cambridge CB2 8EA, United Kingdom

One Liberty Plaza, 20th Floor, New York, NY 10006, USA

477 Williamstown Road, Port Melbourne, VIC 3207, Australia

314–321, 3rd Floor, Plot 3, Splendor Forum, Jasola District Centre, New Delhi – 110025, India

103 Penang Road, #05–06/07, Visioncrest Commercial, Singapore 238467

Cambridge University Press is part of Cambridge University Press & Assessment, a department of the University of Cambridge.

We share the University's mission to contribute to society through the pursuit of education, learning and research at the highest international levels of excellence.

www.cambridge.org
Information on this title: www.cambridge.org/9781009491945

DOI: 10.1017/9781009491952

© Ryan M. Nefdt 2025

This publication is in copyright. Subject to statutory exception and to the provisions of relevant collective licensing agreements, no reproduction of any part may take place without the written permission of Cambridge University Press & Assessment.

When citing this work, please include a reference to the DOI 10.1017/9781009491952

First published 2025

*A catalogue record for this publication is available from the British Library*

ISBN 978-1-009-49194-5 Hardback
ISBN 978-1-009-49196-9 Paperback
ISSN 2517-7273 (online)
ISSN 2517-7265 (print)

Cambridge University Press & Assessment has no responsibility for the persistence or accuracy of URLs for external or third-party internet websites referred to in this publication and does not guarantee that any content on such websites is, or will remain, accurate or appropriate.

For EU product safety concerns, contact us at Calle de José Abascal, 56, 1°, 28003 Madrid, Spain, or email eugpsr@cambridge.org

# The Philosophy of Linguistics

Elements in the Philosophy of Science

DOI: 10.1017/9781009491952
First published online: July 2025

Ryan M. Nefdt
*University of Cape Town and University of Bristol*

**Author for correspondence:** Ryan M. Nefdt, ryan.nefdt@uct.ac.za

**Abstract:** The philosophy of linguistics reflects on multiple scientific disciplines aimed at the understanding of one of the most fundamental aspects of human existence, our ability to produce and comprehend natural language. Linguistics, viewed as a science, has a long history but it was the advent of the formal (and computational) revolution in cognitive science that established the field as both scientifically and philosophically appealing. In this Element, the topic will be approached as a means for understanding larger issues in the philosophy of science more generally.

**Keywords:** philosophy of science, philosophy of linguistics, scientific modelling, infinite generalisation, large language models

© Ryan M. Nefdt 2025

ISBNs: 9781009491945 (HB), 9781009491969 (PB), 9781009491952 (OC)
ISSNs: 2517-7273 (online), 2517-7265 (print)

# Contents

1 Introduction 1

2 The 'Science' of Linguistics 2

3 Conduits to the General Philosophy of Science 18

4 Case Study I: Infinite Generalisation in Linguistics 40

5 Case Study II: Language Models and Linguistic Theory 48

6 The Philosophy of Linguistic Subfields and Future Prospects 55

References 62

# 1 Introduction

Naive reflections on the scientific enterprise might encourage a Tolstoyian impression that 'all natural sciences are alike while all special sciences are special in their own way'. The first clause seems to be a loose generalisation at best. Physics, chemistry, and biology might share the label of 'natural science' but they can differ in their mathematical modelling, methodological assumptions, and scientific targets. The second clause is probably true but largely unilluminating by itself. However, despite the limitations of this sort of thinking, the philosophy of the natural sciences has conditioned how we analyse and investigate the special sciences in numerous ways. Here physics takes the position as a canonical science, with most historical and philosophical work focusing on its development over centuries. Scholars within the special sciences often motivate their 'scientific status' by reference (or reverence for) to physical entities.[1] For instance, there have been genuine questions concerning why anything exists besides physics, if indeed explanations in physical theory are basic and fundamental descriptions of reality (Loewer, 2009). Defences of the explanatory autonomy or worth of the special sciences has been vigorous in response (Dennett, 1981; Fodor, 1974; Kincaid, 2008; Ladyman & Ross, 2007).

In this *Element*, we approach the philosophy of linguistics from the perspective of the philosophy of science. Linguistics, at least since the middle of the previous century, has emerged as a particularly interesting and central special science. A number of considerations have led to this curious position. Firstly, human language has always commanded a certain mythological status among our various other natural endowments. Aristotle ruminated on it, the Stoics formalised it, and Darwin (and Wallace) gave it a central role in our evolutionary success story as a species (see Alter, 1999; Bickerton, 2014). As a science, linguistics has contorted its way from anthropological to mathematical to psychological foundations in the past century and a half alone. Add to this the fact that linguistics, and Noam Chomsky in particular, championed the cognitive revolution in North America, liberating the concept of 'mind' from more circumscribed behavouristic models (Bever, 2021; Chomsky, 1959b; Miller, 2003). Rather than defend or motivate why linguistics should be considered a science or how such a perspective might shed light on the study of language (Nefdt, 2023a), here we attempt the reverse, by asking what developments in the philosophy of linguistics can offer to the philosophy of science itself. Many

---

[1] Of course, mathematical credentials are also often proclaimed with a similar effect. And in turn mathematics holds a special place within the natural sciences as well, see Wigner (1960).

of our in-depth case studies and examples are drawn from syntax, where much of the philosophical reflection and scientific status of the field has been located over the years. Nevertheless, semantics, pragmatics, and numerous other subfields of linguistics are discussed throughout in order to achieve the requisite generality and coverage. Despite this, a perceptible imbalance persists.

Thus, in Section 2, we shadow some of the literature in asking what kind of science linguistics is. This decades-long polemic implicates various issues in the philosophy of the special sciences, including ontology, how to interpret formal models, and whether non-reductive unification is possible. However, the discussion will take the form of connections or insights that philosophical issues in linguistic theory might present for other subfields of the philosophy of science, including cognitive science (Section 2.1), social science (Section 2.2), and the biological sciences (Section 2.3). In Section 3, we explore general issues in the philosophy of science that take special shape within linguistic theory. Here, naturalism, scientific revolutions, theory change, the nature of explanation, and the realism debate are among the central topics. Section 4 puts some of these concepts to work in a case study from the philosophy of linguistics, namely infinite generalisation. I suggest that the philosophy of science can learn certain lessons, both positive and negative, from this mesmerising question in linguistics. Next, Section 5 presents another case study, this time on scientific modelling, by delving into the very pertinent debate concerning the relevance of large language models (LLMs) to linguistic theory. We outline some reasons for and against taking LLMs theoretically seriously, and suggest that scientific modelling questions should feature in the discussion more prominently. Lastly, in Section 6, we briefly conclude by mentioning many sometimes forgotten subfields of linguistics, drawing necessary attention to their worth in the philosophy of science before offering some thoughts on the future of the philosophy of linguistics as a metascientific enterprise.

## 2 The 'Science' of Linguistics

Much ink has been spilled not so much over whether linguistics is a science but rather over what kind of science linguistics is.[2] The assumption is that by providing an appropriate disciplinary base, we can glean the methodological and ontological consequences of linguistic theory.[3] For example, if we follow Katz (1981) and Postal (2009) in their Platonist interpretation of the science

---

[2] This, of course, does not mean that the issue of whether linguistics is a science in the first place has not been addressed in the literature. See Jacobson (1999), Sapir (1929), and Yngve (1996).

[3] This assumption is questionable. For instance, accepting that linguistics is biology does not circumscribe the exact targets or methodology of the field.

of linguistics, then languages are *non*-spatio-temporal abstract objects, and linguistics itself is a formal science akin to logic or mathematics. Although Postal (2009) offers this as a description of the field, accepting it would have revisionary ramifications across the board. For example, the question of the evolution of language in the natural world will have to consider how humans interact with non-natural entities, similarly for acquisition. What becomes more important are the components of formal grammars and 'proofs' about language structure. Thus, instead of offering various candidates for the scientific remit of linguistics as has been conventional practice in the past, I will describe each possible disciplinary home for linguistic theory within its parent philosophy of science. The idea is to evaluate to what extent the philosophy of linguistics dovetails with other philosophies of science and also whether any cross-pollination has proven fruitful to some extent.

## 2.1 Linguistics and the Philosophy of Cognitive Science

A dominant approach in the philosophy of linguistics views linguistics as a cognitive science. There are at least two distinct ways to understand this general position. The first is by far the most prominent within philosophical circles. Generative grammar has inspired a particular way to study the mind and language by introducing the cognitive scientific approach to linguistics.[4] Chomsky (1957, 1965) set the philosophico-linguistic agenda (Freidin, 2013). Subsequent work laid further foundations in the philosophy of linguistics (Chomsky, 1966, 1986, 2000). The second approach under considerations rejects many of the core tenets of the first while retaining the general aim of studying language via studying cognition. Both approaches provide unique insights into profound philosophical questions at the heart of the cognitive sciences.[5] Before we can appreciate either position, we need to discuss the core tenets of generative linguistics and the philosophical consequences of the view.

### 2.1.1 Generative Grammar, I-Language, and Acquisition

The contributions of generative grammar to the philosophy of cognitive science are twofold. The first draws from advances in proof theory and logic in the early twentieth century to establish a mathematical basis for describing various properties of natural language. The specific formalisms were drawn in part

---

[4] It has even spawned its own *sub*field of the philosophy of science, namely the philosophy of generative grammar (Hinzen & Sheehan, 2015; Ludlow, 2011; Rey, 2020).

[5] I haven't defined the 'cognitive sciences' here but roughly the term refers to an ever-changing group of disciplines directed at the study of the mind. These have included cybernetics, artificial intelligence, linguistics, philosophy, and cognitive psychology, among others. See Gardner (1985b) and Frankish and Ramsey (2012).

from Emil Post's work on production systems with inspiration from Nelson Goodman, Alan Turing, and Rudolf Carnap, among others (see Lobina, 2017; Pullum, 2011; Tomalin, 2006). The idea that generativists were trying to capture was couched in a central insight from Wilhelm von Humboldt that language 'makes infinite use of finite means' (Chomsky, 1965; von Humboldt, 1836). Specifically, the mechanism tying the finite vocabulary to the (potentially) infinite formal languages it generates is the role of the grammar $G$. Firstly, we need to split the vocabulary into two sets: (1) a set of terminal elements $V_T$ (which you can think of as words), and (2) a set of non-terminals (or syntactic categories) $V_N$. The total vocabulary is then given by the disjoint union of $V_T$ and $V_N$, with $V^*$ denoting the set of all possible strings of vocabulary elements (Levelt, 2008). From this, $G$ contains two further components: (3) a set of production rules $P$, and (4) a unique start symbol $S$. Thus, $G = (V_T, V_N, P, S)$.[6] $P$ licenses the replacement of strings from $V^+$ with strings from $V^*$ or, following Chomsky's $[\Sigma, F]$ grammars (Chomsky, 1957), strings from $V^+$ can be rewritten as strings from $V^*$. $S$ is a subset or specific element of $V_N$. The task is then to derive the strings (i.e. patterns) of natural language from the combinatorics of the alphabet and a finite set of rules $P$ or the grammar $G$ starting from $S$.

The next aspect of innovation involves where to place this generative grammar or computational procedure. Again, platonists might assume it is describing some abstract pattern or object, but Chomsky opted for a different path to instantiation. Languages, on this view, are internalised systems emanating from an innate set of universal linguistic principles (Universal Grammar or UG), called I-language. Chomsky (1991a, 9) states that 'there can be little doubt that knowledge in a language involves internal representation of a generative procedure'. Specifically, an I-language is an *internal, individual, intensional* mental/brain state of an individual language cogniser (at the appropriate 'level of abstraction'). *Internal* means internal to the language user or intracranial, *individual* pertains to the fact that the grammar does not consider relational facts outside of the individual speaker, and lastly *intensional* marks the functional or generative procedure for getting at the expressions of a language as opposed to the external set containing all such expressions. In other words, there is a generative system (or I-language) inside the minds of human language users (at some level of abstraction).

The two innovations combined offered not only an existence proof of the power of computational theory of mind hypothesis (which underpinned the cognitive revolution of the mid-twentieth century) but also a fulcrum through

---

[6] You can think of the production rules as ordered pairs of strings (from $V$) or the Cartesian product of $V^+$ (a set which excludes the null string) and $V^*$ ($V^+ \times V^*$).

which the other cognitive sciences could pursue related computational goals (Miller, 2003). Gardner (1985a) chronicles the origins of the field of cognitive science and highlights Chomsky's role in operationalising the idea that language can be mathematically characterised with formal precision (allegedly showing the inadequacies of Shannon's information-theoretic model along the way, in Chomsky (1956)).

What is a grammar in scientific terms? According to Chomsky (1995a), it is a scientific theory of the state of the language faculty.[7] Others have disputed the status attributed to grammars. Tiede and Stout (2010) argue that generative grammars are more akin to formal models, while Nefdt (2016) compares them to scientific models. The central issue that drives this particular discussion is the role of linguistic infinity. Basically, generative grammars seem to imply that natural languages are discretely infinity, and have a countably infinite capacity or output. Chomsky calls this the 'basic property' of language or 'each language provides an unbounded array of hierarchically structured expressions that receive interpretations at two interfaces, sensorimotor for externalization and conceptual-intentional for mental processes' (Chomsky, 2013, 647). 'Unbounded' is standardly interpreted as 'countably infinite'. The modelling perspective offers possibilities, drawn from the philosophy of science, to explain unboundedness in ways that do not commit the target system to an actual infinity of expressions, structures, or output. For instance, scientific modelling allows for non-veridical, indirect representation (Godfrey-Smith, 2006; Weisberg, 2007). Models act as surrogate devices for the representation of a given target system (Hughes, 1997). This means that if the model is committed to an infinite output qua generative grammar, the target system (in this case natural language) could still be finite. Nefdt (2019a) shows various cases, from population genetics to physics, in which infinity is a mechanism for 'smoothing mathematics' and enabling generalisation rather than an actual property of the system under consideration (see also Savitch (1993) for the computer science case). Linguistics brings out the general issue of how to interpret artefacts of models to the fore in numerous ways. We will return to the issue of infinity in Section 4.

Under this interpretation of linguistics as a cognitive science, language acquisition takes on central significance. Linguists call this condition 'explanatory adequacy' since it aims to go beyond the description of the data to explaining the underlying principles or mechanisms behind the maturation of language in the human mind. If the language system is internal to language users, then explaining how this process unfolds inside the human mind is of

---

[7] The theory of the initial state with which we are all born is a 'universal grammar'.

paramount importance. There are a number of components, endorsed to differing degrees over various epochs of generative grammar, each of which has garnered philosophical responses.

The project of unearthing the acquisition process has two essential cogs. The first involves a reflection on the data. Linguists generally accept that the data to which children are exposed is impoverished in some way. The details of this protracted debate expose a number of interesting arguments for the philosophy of science in general. For instance, the 'poverty of stimulus' argument holds that mastery of the complexity of natural language is impossible within a short period (the so-called critical period which caps at puberty) without strong innate biases (Chomsky, 1955/1975, 1965; Gold, 1967). The literature is mostly devoted to balancing the various elements of this acquisition quandary, that is, how does a young child acquire a complex language within a short period and without sufficient explicit instruction?

Pearl (2021) provides an excellent overview article in which she separates these elements (speed, quantity, quality, etc.) in terms of their argumentative (and evidential) force. The issues are subtle. They concern not only the amount of data but also the quality, rapidity of acquisition, and extent of external factors in the process. For instance, generative linguists have often focused on the kinds of evidence that are lacking in first language acquisition (or L1). Here, the idea is that children generally do not receive negative evidence concerning unlicensed or illegitimate constructions (Marcus, 1993). Rey (2020) calls these kinds of data 'WhyNots', which are like windows into competence. In other words, given a minimal pair of sentences A and B, children are unlikely to make certain kinds of errors despite lack of exposition to negative evidence concerning those errors. In fact, they seem to opt for A even when B might be the 'simpler' inductive hypothesis or one that's more of an intuitively plausible alternative. Generative linguists have taken such data to indicate a certain kind of innate, structure sensitivity on the part of L1 learners (Chomsky, 1986; Moro, 2016). Specifically, the kinds of structures needed are argued to be the discrete, hierarchical, recursive structures conducive to generative grammars. Others, such as Pullum and Scholz (2002), argue that multiple factors such as prosody are usually omitted from the relevant data under consideration. The 'logical problem of language acquisition', as it is sometimes called, has also received treatment along stochastic lines. Chater et al. (2015), and Hsu and Chater (2010) use minimum description length (MDL) as a metric to estimate the amount of data needed for language learning and argue that discrete grammars can be statistically reinterpreted without strong innate biases. The MDL belongs to probabilistic models with a built in simplicity principle meant to be extracted from general cognition. The idea is that the (idealised) learner is tasked with deciding between grammars based on varying degrees of simplicity in encoding (or compression). Simpler grammars are easier to learn and MDL provides a way of specifying simplicity, that is, as a grammar that minimises the total encoding length.[8] This possibility promises a marriage between formal language theory and statistical approaches to learning. Dupre (2024) insists that such a marriage would be on the rocks from the beginning, as the competence model of grammar espoused by generative linguistics is fundamentally about different things to the performance-based stochastic models (his target is Yang and Piantadosi (2022)'s grammar induction model).

It is profoundly underappreciated just how much this particular debate in linguistic theory has to offer the philosophy of science. Not only does it resurrect the issue of rationalist versus empiricist foundations of science but it also offers insights into the nature versus nurture debate, discrete versus continuous representations in science, and even the epistemology of evidence.

Lastly, a core aspect of the generative approach to linguistics as cognitive science is the relative cognitive isolation of the competence model. Grammar or I-language is a modular and domain-specific system on this account. This move not only separates competence from performance but also from general cognition (Allott & Smith, 2021; Chomsky, 2017; Fodor, 1983; Pylyshyn, 1984). The idea that there are specific areas of the brain devoted exclusively to language (or narrow syntax such as BA 44[9]) has a long history dating back to early work on language disorders (see Baggio, 2022). In a recent review, Tuckute, Kanwisher, and Fedorenko (2024, 281) claim, among other things, that 'the brain's language system constitutes a distinct component of the mind and brain that is specific for language processing, separable from other cognitive systems, and relatively functionally homogeneous across its regions'. However, the nature of this system is still contested. Of course, matters of neurobiology are pertinent, as we will see in Sections 2.3 and 6, but much of the cognitive scientific perspective seems to live at the computational level in Marr (1982)'s hierarchy of cognitive information processing systems, at which the computational problem is stated (what is being computed and why?).[10] Neurolinguistics (and psycholinguistics) by contrast seems to work at the algorithmic and implementation levels wherein we are concerned more with mechanistic explanations.

---

[8] There are natural Bayesian extensions of MDL; see Hsu, Chater, and Vitányi (2011).
[9] See Friederici et al. (2017) for more details.
[10] But see Mallory (2024) for a discussion of the appropriateness of the Marrian analysis of generative linguistics in the first place.

Before we move on to a discussion of how the philosophy of social science might be enriched by considering aspects of the philosophy of linguistics, we should discuss the other (rival) interpretation of the cognitive significance of language.

### 2.1.2 Cognitive Linguistics and Domain-Generality

The place of linguistic theory within the classical cognitive scientific revolution is unassailable. In some ways, linguistics paved the way for the computational theory of mind both mathematically and in terms of its connections to psychological reality. Its insights inspired AI, robotics, and the philosophy of mind. However, computationalism, and linguistics under generative grammar, also involves a number of strong tenets. Many of these tenets have been jettisoned in subsequent movements within cognitive science. Sinha (2010) argues that within the second wave of cognitive science, both the centrality of computationalism and linguistics itself come under simultaneous attack.

What does this alternative picture look like? For one thing, as previously mentioned, the early proof-theoretic results of generative grammar served to operationalise computationalism itself (Chomsky, 1956, 1959a). But they also established a direct link between syntax and linguistic cognition. This connection (or identification) allows linguistics to determine a *sui generis* path, replete with cognitive isolation (and/or modularity) and biological uniqueness. The idea seemed to be very much couched in terms of what (cognitive) science could learn from linguistics rather than the other way around. For example, claims as to the 'generative grammar' within the immune system (Jerne, 1985) or the nature of recursion in cognition (Lobina, 2017) drew from the work of linguistics and formal language theory. The direction of fit was very much how language could help us discover other features of the mind, with a proviso that its uniqueness might block certain analogies. Specifically for Chomsky (1955/1975, 160), '[l]inguistics is simply that part of psychology that is concerned with one specific class of steady states, the cognitive structures that are employed in speaking and understanding'. Mukerji (2022) presents a strong case for using generative linguistics as a tunnel towards a core generative principle (he calls 'Principle G') responsible for language, music, and arithmetic cognition in humans.

However, reversing the order of explanation results in what Lakoff (1991, 54) calls the 'cognitive commitment' or the idea that one should 'make one's account of human language accord with what is generally known about the mind and brain from disciplines other than linguistics'. In other words, ask not what linguistics can do for cognition but what cognition can do for linguistics!

Here the specialness of language is challenged, especially that of syntax. In fact, in most cognitive linguistic settings, the autonomy of syntax (a cornerstone of generative linguistics) is rejected in favour of the inseparability of form and meaning, syntax and semantics. Semantics, the study of meaning, is investigated under this paradigm with more fervour. According to certain hypotheses, such as the much-discussed Hauser, Chomsky, and Fitch (2002) (published in *Science*), syntax is part of the narrow language faculty, while semantics is likely part of an older, more general (or 'domain-general') system. But meaning has always been central to the philosophy of language, and philosophers interested in linguistics alike. As previously mentioned, the focus on form or syntax has informed much of the proof-theoretic or formal underpinnings of linguistics and computationalism in cognitive science. Within cognitive linguistics, computationalism is similarly questioned. M. Johnson and Lakoff (2002) argue that cognitive linguistics actually requires some notion of embodied cognition in which cognition is not identified with intracranial states of individual cognisers but includes our motor movements and environmental effects into the cognitive process. Some philosophers of science have equally embraced embodied versions of cognition to explain the minds of non-human animals, and the evolution of a more comparative concept of consciousness (Godfrey-Smith, 2016). If language is no longer a special brain module, isolated and domain-specific, then it is susceptible to influence of and explanatory integration with other systems.

The kind of science of linguistics that is produced on this view is one with connections to the study of semiotic systems, metaphor, and social pragmatics. For example, work on the phenomenon of conceptual metaphor stemming from Lakoff and Johnson (1999) incorporated experimental psychological methodology, linguistic analysis, and social studies into a coherent whole. It seems indisputable that metaphors can shape the way we think about events, abstract objects, and experiences and thereby condition our actions. Take the now-famous case of arguments. Reasons are considered indefensible, critiques hit their targets, we focus on weak points, and so on. We aim to win them at all costs. This kind of metaphor shapes how we experience arguments and how we act when we are in them.

According to Lakoff and Johnson (1980), this phenomenon is ubiquitous. By using one concept to understand or experience another, it can reveal conceptual components while hiding other aspects. They argue that 'our ordinary conceptual system, in terms of which we both think and act, is fundamentally metaphorical in nature' (Lakoff & Johnson, 1980). There is a logic to conceptual metaphors, which allows for inferences that exploit, to use a term from Sellars (1953), 'word to world relations'.

In Nefdt (2020b), the shifting role of linguistics and the centrality of language within the newly reformed cognitive science is assessed. There, it is argued that linguistics still has a central role to play despite the renewed role of cognitive psychology, which couldn't take its rightful place on the throne initially due to the tyrannical reign of behaviourism (or so the orthodox story goes anyway).

One of the most prominent expressions of cognitive linguistics today is construction grammar. Construction grammarians reject what they describe as 'distributional analysis', which rests on a 'building block model of grammar'. They worry that this approach is not in sync with the typological realities of languages across the world (Croft, 2001). We will return to typology briefly and its neglected role in linguistic theory and the philosophy thereof in Section 6.

Constructions themselves are usually taken to be pairings of form and meaning such that there are 'slots' for saturation of individual constructions such as certain semi-flexible idioms like *X is more Adj than Y*. Grammar on this view is based not on minimal units of composition but the saturation (and learning) of complex entities taken to be basic units of grammar. Goldberg (2015) goes as far as to argue against the principle of compositionality (assumed in formal semantics) which adheres to the building block model of grammar or the view that sentential meaning is built up from the meanings of smaller units (and their syntax).

Despite appearances of opposition, Jackendoff (2002, 2018) describes a comprehensive theoretical framework for incorporating elements of generative linguistics (minus the centrality of syntax), construction grammar, and conceptual semantics. His account, called 'the parallel architecture', also aims to build bridges to psycholinguistics and neuroscience.

The basic idea of cognitive linguistics, and the potential paradigm shift with which it comes, opens the door to the possibility that motor cognition, social cognition, metaphor systems, and so much more can shape how we theorise about natural language. In principle, cognitive linguistics and construction grammar, and so on are more amenable to the techniques of distributional semantics (a statistical study of meaning) and computational linguistics, given its domain-generality. In other words, if language learning is not a special purpose mechanism but a general procedure, then general (even statistical) computational models could exploit this procedure without additional structural biases and possibly even draw from computational models in other cognitive systems such as vision and movement.

## 2.2 Linguistics and the Philosophy of Social Science

It might be strange to those familiar with the philosophy of linguistics to see a section in a book on the topic that includes discussion of its relevance to the philosophy of social science. The kinds of formalisms adopted in linguistics, at least in syntax and semantics, lend themselves to either an individual psychological interpretation or perhaps an abstract ontology. Yalcin (2018, 358), in discussing scientific modelling in formal semantics, states unequivocally that semantics is about 'human linguistic competence—informally, knowledge of meaning'. Glanzberg (2014, 31) concurs with this picture of semantics, insisting that 'it is the state of mind, or the semantic aspect thereof ("knowledge of meaning", understood in the technical sense), that semantics is foremost concerned with modeling'. One criticism of this internalist focus has been the alleged neglect of the communicative dimension of language. Language under certain views is primarily a tool for communication. If there is any kernel of truth in this claim, then linguistics is the study of a distinctly *social* phenomenon.

Many philosophers of language have recognised the central place of communication with their focus on ordinary language, externalism, reference, and conventionalis. Lewis (1975) produced an influential picture of the synthesis of a formal language (or grammar *qua* formal mathematical object) and the conventions of a community of language users. In this sense, a language is both an abstract object and a model of the linguistic patterns of a community of speakers. Still, Lewis' skepticism of the competence model is somewhat of an exception. In addition, his model remains rather abstract.

Within a broadly generative linguistic framework, Labov's work stands out as a canonical treatment of the social aspects of grammar. In fact, his research gave birth to modern sociolinguistics. For example, he discovered fascinating regularities within neglected (or even maligned) dialects of English such as African American Vernacular English (AAVE). Chomskyans generally reject dialect studies at peripheral to linguistic theory as they allegedly concern political vagaries of the so-called external or E-languages. However, Labov showed that studying neglected (and oppressed) dialects of English, such as those spoken by working-class African American communities in the United States, provides a theoretical window into phenomena as diverse as structural variation, language contact, and language change. In Labov (1969), he uncovered a particularly intriguing regularity: that when copula contraction is permitted in Standard English, AAVE allows deletion (and the other way around). Furthermore, when contraction is not possible, then neither is deletion. So you can say *Sane said she going* but not *\*Clever, that's what she*. The AAVE also allows for negative concord, while most other dialects of English do not. Importantly, investigating these grammatical regularities led Labov towards the tools of social science such as peer group recordings. In a sense, Labov showed a fine-grained example of Lewis' synthesis by expanding 'grammar' to account for inter- and intragroup linguistic regularities.[11]

It is hard to conceive of pragmatics without considering sociality and communication. Pragmatics itself is one of the most productive areas of contemporary linguistics (Nefdt, 2024). As early as Grice (1975), the logic behind conversational structure was revealed. This structure is rich and implicates multiple cognitive elements such as 'theory of mind'.[12] The idea is that language involves complex inferential patterns which operate over and above semantic or literal meaning to convey momentary conversation dynamics as well as conventional pragmatic effects. Even the seminal Stalnaker (1978, 2002) formally models the evolution of conversational dynamics as participants interact within an ever-changing common ground of informational states. He defines the common ground as the set of propositions that participants 'mutually accept' for the purposes of conversation, where mutual acceptance involves an iterated acknowledgement between parties of their respective acceptances (A accepts that B accepts that A accepts that ... and so on, *ad infinitum*) (Stalnaker, 2014). Dynamic semantics exploits these dynamics to produce a model of semantic meaning itself as context-change potential (Dekker, 2012; Veltman, 1996).[13]

It might be objected that it is not quite accurate to describe formal pragmatics, the kind of research mentioned in previous paragraph, as a social scientific enterprise for a number of reasons. Here the distinction in Korta and Perry (2008) between 'near side' and 'far side' pragmatics is useful. The former involves work on indexicality and presupposition which implicates interpersonal relations but remains highly formal and abstract. The far side brings a bridge between formal theory and ordinary usage is built. Here they locate Grice's (and J.L Austin's) work, which go beyond saying and approach usage-based accounts of language and communication. As discussed in Keiser (2022), much of pragmatics remains highly idealised and abstract.

A further connection with social science presents itself in the many game-theoretic approaches to pragmatics in formal linguistics. As effective as game theory has been in economics where rationality, social cognition, cooperation, and similar concepts have been extensively modelled, it has also proven a useful

---

[11] Mesthrie and Nefdt (in press) adopt a 'deidealisation' approach (from the philosophy of science) to understand the possible relationship between sociolinguistics and generative theory.

[12] Sperber and Wilson (1995) expand on Grice's pragmatic account to ground it in relevance theory, which is a theory of cognition based on inference and optimality.

[13] Heim (1982) was also influential in the origins of dynamic accounts of semantics.

tool in linguistics. Again, Lewis (1979) explores this game-theoretic approach to communicative presupposition and conversation based on an analogy with baseball scores. In fact, Lewis (1969) established the relevance of game theory, and signalling games in particular, to the study of linguistic conventions. Game-theoretic pragmatics is now a burgeoning field (see Franke, 2013).

The bridge from pragmatics to social science is a short one. Firstly, communicative practices need not assume a monolithic context. Pragmatic phenomena are ubiquitous and are arguable even present at the birth of new languages when language contact is involved (Bickerton, 1975). Horn and Kecskes (2013) discuss two types of social pragmatics. They can be grouped under the terms 'socio-cultural interactional pragmatics' (SCIP) and 'intercultural pragmatics' (IP), respectively. The SCIP broadens the scope of pragmatics to include social and cultural constraints in addition to the linguistic constraints of the mainstream theories. Secondly, it goes beyond the utterance and immediate context level to characterise speaker meaning through the dialogue sequence (sets of utterances) or discourse segment. The IP, on the other hand, incorporates intention-based classical views more centrally while refocusing the field on emergent phenomena within communication. 'In this approach interlocutors are considered as social beings searching for meaning with individual minds embedded in a sociocultural collectivity' (Horn & Kecskes, 2013, 365). This is a form of what the authors call 'sociocognitive pragmatics'.

The social nature of language might go deeper and beyond pragmatics and sociolinguistics. In fact, as far back and Trubetzkoy (1958), phonology and its basic unit the 'phoneme' have been attributed social-normative force. Speech patterns are certainly affected by conformity, gender, pressure, class, and a number of other social factors. Itkonen (2001) investigates the social interpretation of phonetics and phonology while questioning the predominantly mentalist interpretation. In related fashion, in a recent article in *Science*, Fedorenko, Piantadosi, and Gibson (2024) argue that language is optimised for communication and evolved as a means of cultural transmission of knowledge. They marshal information-theoretic evidence in their attempt to dismount the Chomskyan claim that the primary function of language is for thought.

Viewing linguistics as a social science opens up myriad possibilities for the philosophy of social science. Language has been one of the most studied human phenomena. The tools, results, and insights from research into its social components are surely important for social science in general. For instance, pragmatics provides a window into the inferential underpinnings of communication. This in turn gives us insight into linguistic reasoning and social rationality. One example of the possible cross-fertilisation is Kretzschmar (2015). In it, he shows via corpus data that an emergent non-linearity characteristic of market economies is omnipresent in language, namely the Pareto or 80/20 principle in which 80 per cent of wealth is concentrated within 20 per cent of the populace. This dovetails on Zipf's Law, in which the frequency of words in a corpus is inversely proportional to their ranking (i.e. the most common word occurs twice as often as the second most common and three times as often as the next and so on).[14] He claims that the Pareto principle shows up all over the data at various levels and that 'we in language studies can and should make good practical use of the 80/20 Rule on a conceptual basis' (Kretzschmar, 2015, 85). Pareto equilibrium as a tool is also found in game-theoretic pragmatic settings.

It seems clear that the philosophy of linguistics (and linguistics itself) has a lot to offer the philosophy of social science and vice versa. Despite this, the connections are seldom explored within either field. A more prominent connection is between the philosophy of linguistics and the philosophy of biology given the recent focus on language evolution within linguistic theory.

## 2.3 Linguistics and the Philosophy of Biology

Evolutionary concerns have never been far removed from questions of language. Darwin and Wallace reflected on it, with the former favouring an adaptionist, gradualist approach which saw some connections with the origin of other species and their communicative systems (Alter, 2013; Darwin, 1871). In fact, as the story goes, speculation on the origins of language were so rife at some point that in 1866 the Linguistic Society of Paris effectively banned all discussion of it (as did the London Philological Society shortly after in 1872) (see Kenneally, 2007). It took some time for the conversation to not only gather steam again but also eventually resituate theoretical linguistics and its philosophy.

One explicit move in this direction can be found in the seminal work of Eric Lenneberg, especially Lenneberg (1967). Here, Lenneberg lays the foundations for the biological study of language. Specifically, he defined the 'critical period hypothesis'. This hypothesis states that there is a critical period (between two and just before puberty) in which a language learner can acquire their first language. He linked this constrained ability to the process of lateralisation in the brain. But similar critical periods had been noted in other species such as songbirds. The important part of his analysis was its influence on the much later biolinguistics movement. His goal was explicitly to 'reinstate the concept of

---

[14] As noted by Gibson et al. (2021), Zipf's law also pertains to the length of words as more frequent words tend to be shorter. Their general point suggests that pressures on efficiency of communication have actually shaped natural language.

a biological basis of language capacities and to make the specific assumptions so explicit that they may be subjected to empirical tests' (Lenneberg, 1967, vii). This idea of language as either an actual biological organ or a 'natural object' has formed the bedrock of the movement. Anderson and Lightfoot (2002) devote an entire book to developing this biological metaphor into a concrete interpretation of a language growing in the mind of speakers. They conclude that 'there is a biological entity, a finite mental organ, which develops in children along one of a number of paths.... The language organ that emerges, the grammar, is represented in the brain and plays a central role in the person's use of language' (Anderson & Lightfoot, 2002, 22).

There are a number of components to this picture of the biology of language. Some such as modularity, innateness, poverty of stimulus, and grammar are inherited from the erstwhile generative programme, while others such as optimality, evolvability, and neurological structure are taken from a mixture of Chomsky (1995b) and later work by many others. For example, Bever (2013) reviews the development of the biolinguistics programme from its inception to the (then) present day. He specifically investigates in what sense language viewed as an organ corresponds with other senses of the word in biology. This brings up the work on the genetics of language. At some point, scholars were optimistic about the possibility of a FoxP2 or 'the language gene' (Gopnik, 1990). Among other things, such work offers a genuine path to the innateness characteristic of biolinguistics. As Bever (2013, 401) notes, 'the usual method (in principle) is to isolate a particular genetic abnormality and relate it to the selective sparing or selective impairment of language ability, thereby making more specific the claim that language is "innate".

However, the hype outstripped the hypothesis. While the idea of a grammar gene offered biological grounding for the field, further studies proved that the FoxP2 gene was not human specific, nor did defects associated with it manifest purely linguistically (Sampson, 2005; Vargha-Khadem, 2005). Murphy (2012) insists that the posit is more revealing of linguistic performance than competence in any case. Nevertheless, the work of Lenneberg, Chomsky, Bever, and others aimed at uncovering the biological substrate of linguistics and resetting the agenda. It can be seen as what Boeckx and Grohmann (2007), in their launching editorial of the journal *Biolinguistics*, call 'the strong sense' of the term. This idea involves the use of insights from biology as a driving force for linguistic theory. The weaker sense does not constrain generative linguistic theory directly as the minimalist programme would suggest or work on the genetics or neurobiology of language might do. Murphy (2012) approaches this topic from numerous angles. His exposition of biolinguistics and its relation to philosophy (chapter 1) highlights a number of connected strains, both in terms of weak and strong versions. In it, he makes a potent case for the modularity of language, the organic interpretation of I-language (quite literally growing in the brain), and the ancillarity of communication to language. It certainly seems clear from Murphy (2012) (and Chomsky, 2000) that the traditional philosophy of language offers little in terms of correspondence with human biology.

One interesting aspect of this model of biolinguistics (in either the weak or strong form) is its relative isolation from standard biological theory. The most prominent departure presents itself within the saltational account of the evolution of language. Minimalism is the view that the basic mechanism of language (or narrow syntax) is the operation of Merge. It is a bottom-up structure building operation that takes two syntactic objects and combines them with the projected label of the head. Merge acts as our explanans towards the goal of modelling the explanandum or 'basic property' of language: 'each language provides an unbounded array of hierarchically structured expressions that receive interpretations at two interfaces, sensorimotor for externalization and conceptual-intentional for mental processes' (Chomsky, 2013, 647). This latter property is usually connected to the linguistic infinity postulate (which we will encounter in Section 4). From an evolutionary perspective, the task of explaining the origins of language becomes a task of explaining the origins of Merge. The biolinguistic story goes something like this:

> At some time in the very recent past, apparently sometime before 80,000 years ago, if we can judge from associated symbolic proxies, individuals in a small group of hominids in East Africa underwent a minor biological change that provided the operation Merge—an operation that takes human concepts as computational atoms and yields structured expressions that, systematically interpreted by the conceptual system, provide a rich language of thought. (Berwick & Chomsky, 2016, 87)

What results is an explosion of intelligence or a macromutation which set human beings on a path towards cognitive dominance on the planet. But whither natural selection, deep homologies, the tapestry of cross-species adaptations when it comes to language? Many of the standard accounts in evolutionary biology seem to be missing in this project. For instance, on this account, communication is an exaptation. The central evolutionary nexus is the language faculty, represented by its proxy Merge, meaning that 'communication is merely a possible function of the language faculty, and cannot be equated with it' (Friederici et al., 2017, 713).

But of course, every part of the puzzle is up for grabs. One could question the timeline with Dediu and Levinson (2013) and Steedman (2017). One could question the uniqueness with Everett (2017) or the trigger with Bickerton

(2014), who adds a niche construction theory on top of the Merge saltation account. Views like Tomasello (2008) embrace the communicative and cross-species approach while introducing pragmatics and gesture to the evolutionary fold. Even staying within generative grammar, one could opt for a gradualist (more Darwinian) approach with Progovac (2015, 2016). Or one could even question whether Merge's emergence was a single step or broken up into multiple stages (Martins & Boeckx, 2019). There are more stark departures and various variations on the theme (Nefdt, 2023b). The situation perhaps makes one sympathetic to the French academy's erstwhile injunction.

Many philosophers, and cognitive linguists, reject the demotion of communication in language evolution. Millikan's work stands out as both philosophically rich and directly inspired by biological theory. Her view draws from an overarching philosophical framework called teleosemantics. Teleosemantics covers more than just linguistic entities like sentences and words but the nature of representation itself. In contrast to truth conditional accounts of meaning and representation, '[a]ccording to the teleosemantic program, representations are states whose biological function is to guide behavior in ways appropriate to such-and-such conditions' (Papineau, 2016, 97). The states in question are usually mental states like beliefs, desires, perceptions, and so on. However, Shea (2018) presents a compelling telesemantic account of subpersonal representational states. Millikan directs her attention towards linguistic conventions and a uniquely biological account of words.

Millikan's account starts with the concept of proper function (Millikan, 1984, 2005). The proper function of something is basically what that thing is supposed to do (hence the 'telos'). The function of my heart is to pump blood, the function of my pen is to write. The rabbit's stomping of its back legs is a social indication of danger to other rabbits like the bee's famous waggle dance informs other bees of the direction and distance of food sources. Something may have a proper function even if it malfunctions in some way. Millikan's extended this to linguistic representations as biological categories with proper functions. Two subsystems comprise this representation system: (1) producer systems, and (2) the consumer systems. Take the bees again. The producer's function (to produce a dance) and the consumer's function (to retrieve information on food sources) contribute to the function of the entire system (the colony's survival). This system only works in normal conditions (which give us the content of the representation). With this in place, and a modification of Lewis (1969, 1975), the resulting view of language is one in which communication is the primary function and language is 'a set of speaker-hearer interactions forming lineages roughly in the biological sense' (Millikan, 2003). Words on such a view are like species identifiable by their lineages (see Richard (2019) for a similar view of 'meanings' of words as biological species).

The intervention of cultural norms and conventions on linguistic evolution is taken up in later work as well. Dennett (2017) incorporates work on genes and memes to produce a mimetic account of the proliferation and evolution of language, with words at the centre. Planer and Sterelny (2021) start from the point of the messiness of biological kinds to develop a multi-causal, empirically rich picture of the evolution of language, divorced in some ways from the formalised models of linguistic theory. Nefdt (2023b) invites complexity science and biological systems theory into the discussion of language evolution while attempting to maintain some connections with more minimalist accounts of biolinguistics.

Once again, the fruitful collaboration between biology and linguistics offers many avenues of insight. In turn, the respective philosophies of these subjects are intertwined in terms of their commingling of culture, genetics, and cognition. And we have not considered the role neurobiology at all so far. Of course, the field has been marshalled to support generative linguistics (Berwick, 2013; Friederici et al., 2017; Marantz, 2005; Murphy, 2023) and also to question the connection between linguistic theory and biology (Baggio, 2020; Poeppel & Embick, 2005). We will return to neurolinguistics in Section 6 as it possibly offers the most promising grounding for biolinguistics and linguistics in general. Another promising line is pursued Balari and colleagues. Balari et al. (2013) offer a palaeontological perspective while Balari and González (2013); Balari and Lorenzo (2009) apply evo-devo thinking to biolinguistics.

From the perspective of the philosophy of biology, two contrasts seem to unfold: either language is truly biologically unique and thus pushes biology itself to adapt to new tools of exploration or philosophers of linguistics (and theoretical linguists) have not explored the full range of options from biology to account.

## 3 Conduits to the General Philosophy of Science

In this section, the focus will be on how issues within the philosophy of linguistics either exemplify, offer case studies, or challenge well-established conceptions in the general philosophy of science. Each subsection constitutes a vignette or snapshot of a general problem or topic and how it has manifested within the philosophy of linguistics. Section 4 is more detailed, given its centrality in theoretical linguistics and importance in the philosophy of science.

## 3.1 Scientific Revolutions

The nature of scientific revolutions and the concomitant sociology of science have been enduring topics within the philosophy of science since, at least, Kuhn's magnum opus: *The Structure of Scientific Revolutions* (Kuhn, 1962). The basic picture is this: normal science operates with relative progress and consistency under a broad paradigm or governing framework, that is, until the problems mount for the paradigm over time, resulting in a crisis. While 'rationality' might be attributable to normal science, crises are capricious and often driven by politics, historical accidents, and personal factors. In turn, crises are the only revolutionary mechanisms which can dismount entrenched views and theories. Despite their essential function, revolutions remain disorderly and chaotic. Indeed, Kuhn suggests that this might be a consequence of the fact that scientific evidence itself is under scrutiny in a revolution.

Many, if not most, of these components have been challenged in some form or other. For example, it is not clear whether crises precipitate paradigm shifts in every case (Godfrey-Smith, 2003). It is equally unclear whether true incommensurability follows from revolutions as structuralists have pushed for decades (Ladyman, 1998; Worrall, 1989). But as an idealised model, some core elements of the Kuhnian picture remain useful for understanding revolutions (linguistic or otherwise), such as (1) periods of paradigmatic normal science, (2) accumulation of quandaries leading to revolution, (3) the establishment of a new paradigm, partly inspired by performance on those quandaries, and (4) the fact that (3) is driven by non-rational and sociological factors.

### 3.1.1 The Cognitive Revolution

There are at least two avenues for fruitful insights into scientific revolutions that present themselves in the philosophy of linguistics. The first is more obvious than the second. Let's start there. The early work of Noam Chomsky has been regularly described as a scientific revolution. Indeed, it certainly seems true that generative linguistics, along with a few other neighbouring disciplines, ushered in a revolution, in the Kuhnian sense, within cognitive science (Bever, 2021; Miller, 2003). In a *Nature* review, Smith likens Chomsky to Darwin and Descartes. He defends the location of the book review (of Harris' *The Linguistic Wars*, to which we will return) in the following way.

> That a book on the history of linguistics should be reviewed in *Nature* is ultimately due to the fact that Chomsky's work has brought the study of language from the impressionism of the humanities into the scientific fold. (Smith, 1994, 521)

How did he achieve such a lofty goal? Well, the first step was to slay the behaviourist behemoth which came before. Behaviourism, much like logical positivism, restricted psychology or the general study of the mind to observable behavourial dispositions. Chomsky (1959b)'s review of B. F. Skinner's radical behaviourist take on language in *Verbal Behavior* (Skinner, 1957) certainly caused a stir, if not fomented a revolution. In fact, this work and others further afield in cybernetics, AI, and neuroscience could be argued to have created a critical mass of issues for behaviourist models of cognition and language. In addition, mentalistic computationalism offered better accounts of the phenomena which proved to be its predecessor's breaking points. One of Chomsky's critiques was that the stimulus-response model is inadequate as an explanation of the productivity of language (the fact that we can produce a seemingly infinite set of structures from finite rule systems, more in Section 4). Following this, work in many of these areas, especially linguistics, gathered steam and generative linguistics is nothing if not an exciting paradigm in the history of science. What is less clear is the extent to which there was a revolution in linguistics itself.

An early application of Kuhn's picture to linguistics can be found in Percival (1976). In it, he argues that a revolution certainly seems to have occurred in the history of linguistics with the advent of Chomskyan generative grammar. However, he disputes the use of the term 'paradigm' as Kuhn viewed it. Kuhn somewhat controversially requires both that the establishment of a paradigm involves a single scientific genius (like Copernicus or Einstein) and that some sort of scientific consensus ensues from the initial displacing discovery. Chomsky has arguably fulfilled the first requirement but the second is what Percival questions. Of course, he was writing at a time shortly after Chomsky (1957, 1965). Although Newmeyer (1986a), writing some time later, who is much more convinced that a linguistic revolution took place after Chomsky (1957), agrees that the paradigm did not institute a comprehensive power shift. He states, somewhat regally, that 'far from being comfortably seated on the throne after their successful "palace coup"-generativists, as they compete for adherents with linguists of other persuasions, find themselves well outside the walls of the palace' (Newmeyer, 1986a, 2). Of course, again, time told a slightly different story. Generative linguistics did eventually achieve relatively dominant status, particularly within theoretical linguistics. In fact, this was despite the radical changes to the theory over time. As Lappin, Levine, and Johnson (2000) stress with the advent of Minimalism, many linguists followed Chomsky into a 'major paradigm change in the theory of grammar'. However, they argue that unlike analogous paradigm shifts in physics (such as Compton's work on quantum explanations of electron scattering or the move to quantum physics from special relativity and earlier classical mechanics), the transition from Government and Binding (Chomsky, 1981) to Minimalism (Chomsky, 1995b) was 'unscientific' by comparison. Whatever the veracity of their claims, yet another interesting aspect of the philosophy of linguistics emerges, that is, the possibility of paradigms within paradigms. Bickerton (2014) even suggests that linguistics is scientifically unique in that you have people working within different paradigms of the same general programme in the same department![15]

### 3.1.2 Pre-paradigm Science

I think there is a different issue with directly applying Kuhnian paradigm shifting terminology to linguistics. It draws from two independent sources. The first is that it is unclear that there was a distinct pre-generative paradigm prior to Chomsky's intervention and, furthermore, generative grammar itself builds on the work of other formal linguists such as Chomsky's mentor Zellig Harris. The standard historical foil is what has been referred to as 'American Structuralism'. The claim is that linguists operating under its banner rejected universalism, practised taxonomic discovery methods, and borrowed (from behaviourism) an unhealthy scepticism towards unobservable 'mental entities'. Here, Bloomfield is often quoted as expressing the general sentiment.

> Non-linguists (unless they happen to be physicalists) constantly forget that a speaker is making noise, and credit him, instead, with the possession of impalpable 'ideas'. It remains for linguists to show, in detail, that the speaker has no 'ideas', and that the noise is sufficient. (Bloomfield, 1936, 23)

But again, it is not clear to what extent Bloomfield actually espoused the structuralism of de Saussure (which had clear precepts) or whether those who followed his lead in North America could be considered to constitute a unified paradigm. This much is forcefully pushed in both Joseph (1999) and Matthews (2001). In fact, it is not even clear to what extent linguists around Chomsky were not generally supportive of his ideas. Furthermore, Chomsky's transformational grammar drew strongly from the linguistic work of Harris (a structuralist) and the formal work of Post (see Pullum, 2011). Although even formally Chomsky's concept of a generative grammar as a recursive procedure can be traced back to Harris' philosophy of linguistics, as argued by Mallory

---

[15] This is an intriguing possibility and might indeed intimate the potential uniqueness of linguistics as a field. However, similar descriptions seem apt for fields like computer science in which some scholars are working on formal language theory, others on computability, and yet others on deep learning (Rapaport, 2012). In fact, if this characterisation of linguistics proves reasonable it might militate against a Kuhnian analysis in favour of something like a Lakatosian one, where competition fuels progress.

(2023). But does Chomsky's intellectual influences, like Harris, Quine, Bar-Hillel, Jespersen, and others (see Tomalin, 2006), really preclude his status as a revolutionary? This is perhaps another useful example of the limitations of Kuhn's historiography of science.

Nevertheless, the Kuhnian picture pushed in different directions. One prominent strand involved the shift of focus from explaining science as a rational process to understanding it partly (and largely) in terms of its sociology. Here too linguistics is a treasure trove.

### 3.1.3 The Linguistic Wars

In a recent re-edition of his 1993 book, Harris (2021) details a particular episode in the history of linguistics. Early transformational approaches to grammar posited a special kind of structure called 'deep structure' in the architecture of the language faculty. Without too much detail, the basic idea is this: phrase structure rules can account for a number of linguistic structures such as how a verb phrase is composed of a noun phrase and a verb ($VP \rightarrow NP + V$) as in *The woman sings*. But there seems to be an intimate (semantic) relationship between *The woman sings a song* and *A song is sang by the woman*. This correspondence between active and passive constructions in English seems to require a transformation of some sort, one that holds meaning invariant. Initially it was assumed that there is a 'deep structure' underlying the surface structures of sentences (not always the same as the actual output of the sentence) from which different constructions could be derived (via a series of operations like movement and deletion, etc.). In Chomsky (1965), deep structure was a mechanism for the description of formal properties of surface structure (on the assumption of an autonomous syntactic representation). One consequence was that meaning could not be altered with transformations or transformations are semantically invariant. But for an emerging group of radicals called 'generative semanticists', deep structure promised more connections to semantics. For Chomskyans, at this time, transformations were mediators between deep structure and surface structure. For the generative semanticists, transformations were a window into deep structure, which was in turn very close to what logicians considered to be the logical form of the sentence. In their careful take on the same time period and issues, Huck and Goldsmith (1995) differentiate between studying syntax as a distributional system versus studying it as a mediational one. Here, 'distributional' is concerned with formal properties of constituents, while 'mediational' involves the ways in which those properties can be mapped on semantic ones. The semanticists wanted to explore the latter, while the orthodoxy stuck firmly to the former. Scientifically, the issue also involved just how far to stretch the paradigm. Generative semanticists started to model such a vast array of phenomena (from semantics and pragmatics) that Chomskyans considered ill-fit for purpose and an encroachment on the edifice of formal syntax.

From these technicalities, an internecine battle over the heart of generative grammar ensued. This battle went well beyond academic disputes and took on personal force, even resulting in invective hurling at Linguistics Society conferences. Eventually, the generative semanticist camp seemed to have lost the war but many of their core posits were incorporated into later versions of the Chomskyan picture (without due reference, according to Harris) and spawned their own enterprises (such as the cognitive linguistics we encountered in Section 2.1.2). Science can be a messy business. Episodes like these in theoretical linguistics have been documented not only by linguists such as Goldsmith and Laks (2019), Huck and Goldsmith (1995), and Newmeyer (1986b, 1996) but also by, as mentioned, historians of scientific rhetoric like Harris (2021). At the very least, if you are tired of reading about Einstein and Bohr, you might want to learn about Chomsky and Lakoff.

### 3.1.4 Relativism about the Language of Science

Lastly, at least one additionally fascinating strand links Kuhn to theoretical linguistics and that's his relativism. Kuhn's initial view (ignoring later amendments) was quite radical. This is especially the case when one considers the relationship between *pre*-paradigm and *post*-paradigm science. It all revolves around his concept of incommensurability. Kuhn is not entirely consistent but interpreters have discerned at least two important notions of 'incommensurable'.

> First, people in different paradigms will not be able to fully *communicate* with each other; they will use key terms in different ways and in a sense will be speaking slightly different languages. Second, even when communication is possible, people in different paradigms will use different *standards of evidence and argument*. (Godfrey-Smith, 2003, 92)

Now, this notion has caused consternation within the philosophy of science. But a cursory glance at the Sapir-Whorf hypothesis or linguistic determinism in the history and philosophy of linguistics might have set off even more alarm bells. The idea there was that language determines thought (or the limits of thought) and that your language is a product of your environing community. The resulting picture is one of *linguistic* incommensurability. Speaking different languages involves accessing different worldviews which others (outside of your community) simply cannot do. Most scholars today do not accept anything close to this strong thesis, even if recent experimental evidence in cognitive anthropology suggests some divergences in linguistic cognition (Boroditsky, Schmidt, & Phillips, 2003; Deutscher, 2010; Pelletier & Nefdt, 2025; Seuren, 2013). Consider the first sense of incommensurable. In the philosophy of linguistics, this kind of posit would invite a question over bilingualism. If people cannot fully communicate with each other due to incommensurable worlds, then the possibility of a bilingual speaker seems farfetched. But outside of certain geographical locations, bilingualism is rife to the point of ubiquity. Similarly, within scientific discourse, scholars are often fully conversant in multiple paradigms, the philosophy of science itself would be impossible if this were not the case.

The second sense too touches on deep issues in the methodology of linguistics between so-called universalists and particularists. Many formal linguists tend to favour the view that there is some universal (*and* mathematically precise) substrate of language common to all human languages. The influential Principles and Parameters framework exemplified such a possibility (Chomsky, 1981). Other more typologically-minded linguists resist this kind of view and suggest that diversity of linguistic structure is more of a datum than universality (Evans & Levinson, 2009). However, Berwick and Chomsky (2016, 93) insist that 'the appearance of complexity and diversity in a scientific field quite often simply reflects a lack of deeper understanding, a very familiar phenomenon'. Here, there seems to be 'different standards of evidence and argument' at play.[16]

It might turn out that the Kuhnian idea that sciences or scientific theories are like languages is little more than an unhelpful metaphor. But the topic of linguistic relativity certainly charted a well-worn path on incommensurability that we are only beginning to disentangle from the common conception of language now (McWhorter, 2014). Insofar as these (folk) concepts filter through to scientific discourse on theory change and continuity, we would be remiss not to heed the warnings of the philosophy of linguistics here.

## 3.2 Justification and Explanation

Two related and central topics in the philosophy of science are justification and explanation, respectively. There are a number of theories and models of how theories and models work in science. What counts as justification or confirmation is a deeply epistemic suite of questions. What makes a good explanation

---

[16] Another clear example of different standards is the interpretation of 'noise' in the data. Formal linguists tend to abstract over certain patterns that sociolinguistics, and corpus linguistics take to be essential.

cannot be thoroughly divorced from psychology (for Quine (1969), epistemology was just a form of psychology). In many senses, the trajectory of linguistic theory has attempted to borrow the justificatory methods and explanatory status of the natural sciences. The project of naturalism is rarely seen in such splendour across the other special sciences (Chomsky, 2000; Johnson, 2007; Nefdt, 2023a). In one way, this formal rigour and naturalistic aspiration might have inoculated linguistics from the 'theory crisis' emerging in the other special sciences like psychology (Markus & Bringmann, 2021; van Rooij & Baggio, 2020). On the other hand, the empirical credentials of linguistics has often been called into question due to the kinds of idealisations it celebrates, namely rationalistic mathematical ones. Theoretical linguists have also favoured explanation over prediction, which manifests itself in a general disdain for stochastic methods (see Section 5). This conglomeration of factors produces a thick soup of ingredients for the philosophy of science. Unfortunately, the short space afforded here can only facilitate a guide to the literature but hopefully that will suffice as an entreaty to further investigation.

### 3.2.1 UG and Unfalsifiability

Let us start with justification. In the philosophy of science, this issue has generally been approached at a very high level. Somewhat infamously, the logical positivists put forward the criterion of verification as a means of separating the scientific wheat from the pseudo-scientific chaff or rather a criterion determining meaningfulness of scientific statements. For example, to meet the verification principle, one only needs to show that your theory or posit is verifiable *in principle*, that is, the (ideal) conditions under which it can be verified. This still casts out many metaphysical claims (as was the original intention) but operates at a relatively abstract level of justification. But verification proved unstable as a founding principle for scientific endeavour as it could never be fully achieved. Even theories that enjoy relatively comprehensive verification in their predictions could fall with the next prediction failure.

> What this suggests, however, is that a hypothesis can never be completely verified and so the verifiability view needs to be modified. With this in mind, the logical positivists began to shift their emphasis from verification to confirmation. (French, 2014, 79)

But again, issues pertaining to the tribunal of experience or the 'Duhem-Quine' problem challenged confirmation as well as verification, perhaps even more so. Briefly, what exactly is being verified or confirmed since scientific claims usually come with a host of auxiliary hypotheses which connect theory or experiment to evidence? No straightforward answer is available.

In the wake of unconfirmable verifiability or unverifiable confirmation, Popperians and their falsification metric achieved impressive uptake, not only in the philosophy of science but also in the actual sciences themselves.[17] Here the operative criterion is falsifiability. The more precise a theoretical claim, that is, the more easily it can be shown to be wrong, the better it is. As counter-intuitive as this sounds, it proves to be much more efficient at demarcation, a core job description of earlier twentieth-century philosophers of science, than verification. Unscientific claims tend to be vague and difficult to scrutinise. Horoscopes hedge their bets. It turns out that Popper's tool explains why this makes for bad theories. The improvement is palpable for Popper.

> Scientific theories can never be 'justified', or verified. But in spite of this, a hypothesis A can under certain circumstances achieve more than hypothesis B - perhaps because B is contradicted by certain results of observations, and therefore 'falsified' by them, whereas A is not falsified; of perhaps a greater number of predictions can be derived with the help of A than with the help of B. (Popper, 1959, 315)

From this quote, falsifiability can be seen as a positive tool for theory comparison and justification. But since then, despite wholesale rejection of Popper's views, *un*falsiability has been retained as a hammer or an axe to cut down overreaching theories. In fact, the Universal Grammar (UG) postulate itself has been accused of being unfalsiable (Dabrowska, 2015; Evans & Levinson, 2009). The problem is that UG requires the presence of strict linguistic universals, common structures or patterns across the world's languages. But for Dabrowska (2015) and others, positing these kinds of 'invisible' structures makes the theory unfalsifiable. Evans and Levinson (2009, 429) go a step further to claim that '[t]he claims of Universal Grammar, we argue here, are either empirically false, unfalsifiable, or misleading in that they refer to tendencies rather than strict universals'.

The issue, of course, cannot be about invisible or unobservable entities in science. Such creatures pervade every science from the mature to the inchoate. Take linguistic semantics as an example. In the business of aligning expression with their truth-conditional counterparts, semanticists help themselves to a host of theoretical entities from variables (bound or within the scope of quantifiers) to operators hiding beneath the surface of strings. Stanley and Szabó (2000) argue that hidden elements in syntax are necessary to explain quantifier domain restriction in semantics. On this view, NPs have covert argument places for domain restriction in contexts. Invisibility pops up at the interfaces

---

[17] This is apparently especially the case in the medical sciences, see Taran, Adhikari, and Fan (2021).

as well. Approaches like Jackendoff's parallel architecture (Jackendoff, 2002) are structured around multiple generative components linked by bridging principles. Given this structure, gaps between syntax, semantics, and phonology naturally emerge. In other words, there are syntactic elements without phonological overlay (PRO, traces, null determiners, etc.), phonological units without syntax (idioms, constructions), and even possible languages with semantics and limited or no syntax, as argued in Jackendoff and Wittenberg (2014).

It is a big question as to what percolates through from syntax to semantics if we formally treat each domain as a relatum in a homomorphism. In other words, certain posits of syntax or semantics might not be visible to the other formalism or structure. Consider the 'Extended Projection Principle' which operates in languages like English to ensure the existence of subjects whether or not they can receive semantic interpretation (as in '*it* is raining') (Chomsky, 1981). In accordance, many semanticists would treat the mapping between syntax and semantics as somewhat porous, that is, not everything gets an interpretation. However, Jacobson (2012) challenges this assumption in her 'direct compositionality' framework.

> The hypothesis of Direct Compositionality is a simple one: the two systems work in tandem. Each expression that is proven well-formed in the syntax is assigned a meaning by the semantics, and the syntactic rules or principles which prove an expression as well-formed are paired with the semantics which assign the expression a meaning.... It is not only the case that every well-formed sentence has a meaning, but also each local expression ('constituent') within the sentence that the syntax defines as well-formed has a meaning. (Jacobson, 2012, 9)

The positing of covert material and the existence of 'invisible' theoretical entities is commonplace in linguistics as it is elsewhere in the sciences. Recently, Schlenker (2018) takes a step further to argue that evidence from sign languages actually enrich the theoretical landscape by revealing surface realisations of previously covert postulates (in spoken languages). For example, sign language loci (spatial indications of anaphoric content and context) act like variables in different ways. This is fascinating from the perspective of 'Universal Semantics' (the semantic side of UG), since it shows, Scklenker argues, the actual presence of invisible elements from logical form in the language faculty. The idea that some languages make explicit what others assume (structurally) is a cornerstone of cross-linguistic approaches to UG. Minimalism hypothesises deleted copies of lexical items as movement occurs (Chomsky, 1993). Afrikaans can sometimes display multiple copies in the surface syntax, a fact that combined with UG can suggest the existence of copies in a similar fashion to Scklenker's arguments concerning certain sign languages.

The issue seems to come down to different standards of justification. Consider the heated exchange between Chomskyans and a former acolyte Daniel Everett.[18] The tale in a pine-nutshell is that Everett spent time with the isolated Pirahã tribe in Brazil studying their language and culture. There, he deigned to question whether a long-held universal of language, recursive syntax, was indeed present in the language spoken by the people in this region. He concluded that parataxis (sandwiching words together in linear order) sufficed for their needs and careful grammatical analysis revealed no hierarchical or embedded constructions, and even limited access to numerical representation (Everett, 2005, 2012). We're not focused on the veracity of the claim here but rather the response it engendered in terms of justification. Of course, many questioned the empirical claims too (Nevins, Pesetsky, & Rodrigues, 2009), which suggests the auxiliary hypotheses were being questioned.[19]

The response was basically that even if Pirahã did lack recursion on the surface, UG would not be falsified. This is because UG goes deeper than surface syntax. Pirahã is recursive whether it likes it or not! It just doesn't know it. This is why critics have accused generativists of unfalsifiability, since it is not clear what would count as counter-evidence (if language data doesn't do the trick).

### 3.2.2 Galilean Science in Linguistics

It seems that the controversy discussed in previous sections stems from a differing concept of what counts as evidence and justification. UG depends in part on a different kind of confirmation, more rationalist and mathematical. In fact, here Chomskyans draw from a particular interpretation of Galileo. Chomsky offers this take on the philosophy of science behind his rationalism.

> What was striking about Galileo, and what was considered very offensive at that time, was that he dismissed a lot of data; he was willing to say 'Look, if the data refute the theory, the data are probably wrong.' And the data that he threw out were not minor... just said: 'We'll live with the problems and do the mathematics and some day it will be figured out', which is essentially Galileo's attitude towards things flying off the earth. That's pretty much what happened.... And what's true of mathematics is going to be true of everything.... The recognition that that's the way science ought to go if we want understanding, or the way that any kind of rational inquiry ought to go—that

---

[18] This is another story for the sociology of science history books. See Gibson and Poliak (2024) for a volume that covers the controversy and much more, especially chapter 2 by Geoff Pullum.

[19] In fact, there are some interesting unexplored parallels between linguistic fieldwork and natural experiments in science. See Morgan (2013) for some interesting thoughts about natural experiments in social science.

was quite a big step and it had many parts, like the Galilean move towards discarding recalcitrant phenomena... That's all part of the methodology of science. (Chomsky, 2002, 98)

There is a lot to unpack here. Some, like Botha (1982) and Behme (2014), have strongly critiqued this interpretation of Galileo and the position in the philosophy of linguistics espoused by it. Others, such as Boeckx (2010), J. Collins (2023a) and Allott, Lohndal, and Rey (2021a), attempt to reconcile the so-called Galilean style with contemporary philosophy of science. Here the role of abstraction and idealisation is central to the argument. For instance, Allott, Lohndal, and Rey (2021b, 519) argue that 'any serious science must study each system largely in highly idealized isolation from the others'. They compare the competence—performance distinction to the elimination of friction and wind speed from Newton's laws. The idea they push is that models can isolate aspects of systems which are extraneous to those systems. Similarly, a direct application of the scientific modelling literature in the philosophy of science to linguistics is pursued in Nefdt (2016). There, following Weisberg (2007)'s characterisations, generative grammar is linked more to 'minimalist' models than what he calls 'Galilean idealisation', since the former, focused as it is on finding a minimal causal basis for a phenomenon, does not make necessary room for deidealisation.

The important part is that for certain linguists, the mathematical models will trump the empirical data, at least initially (as the Chomsky quote states). For instance, Boeckx (2006, 124) describes the connection between the Galilean style and minimalism as '[s]ometimes, the mathematical results are too beautiful to be untrue, so that it seems justifiable to stick to the theory, while setting aside problematic or even conflicting data'. Indeed, minimalism does privilege simplicity and optimality above other 'extraneous' considerations. Chomsky initially acknowledges that language is a 'messy' biological system (Chomsky, 1995b, 29). However, he goes on, a few lines later, to reinforce the need to adopt minimalist assumptions as a 'working hypothesis' of the basic structure of language, based as they are on simplicity and elegance.

It is unlikely that rationalist versus empiricist philosophy of science will be settled by the philosophy of linguistics, although some were hopeful that it could be (see Katz, 1972). Of course, these kinds of debates don't just touch on justification but the very nature of observation as well. It is well known in the philosophy of science that observation is a theory-laden endeavour. To generative linguists, perceptual or surface observation statements (concerning the grammar of language X) cannot in principle dismount a deeper theoretical statement (and Universal Y) if that deeper statement is supported by mathematical theory. Language, for the former, is an internal mental state not a public collection of sentences or sets of behavioural dispositions. Internal mental states are not observable directly, without the intervention of theory. It seems that mathematical theory is guiding observation in the minimalist programme in a way that opponents deem to be inappropriate. The vocabulary of the philosophy of science can offer a stage setting for these debates without which they seem partisan and irreconcilable. On the flip side, these debates in the philosophy of linguistics offer the philosophy of science more data to access and analyse if due attention is applied. We will return to some of these issues in our first case study, but for now a brief word about a related and long-standing impasse concerning the nature of evidence in linguistics.

### 3.2.3 Linguistic Evidence

There have been many articles and volumes on the issue of linguistic evidence (Cowart, 1997; Featherston, 2007; Phillips & Lasnik, 2003; Schütze, 1996). In some cases, the very scientific status of linguistic theory seemed to be at stake. Contemporary linguistics is more eclectic and welcoming of different methodologies and kinds of evidence. Thus, I believe this issue is largely settled. Nevertheless, it is another case of useful grist for the metascientific mill. So I'll briefly mention some of the key components here.

Firstly, the particular debate is really a debate about how generative linguistic theories proceed. It is impossible to describe 'the methodology' of linguistics *in toto* beyond this circumscription. Sociolinguists use tools from sociology and social science, neurolinguists use fMRIs and CT scans, probabilistic linguistics trades in annotated corpora and Bayesian probabilities, and field-linguists produce surveys and interview people. The list goes on. For a long time, it looked like generative linguists gathered linguistic intuitions as a primary driver of theories. What seemed worse is that they include their own intuitions in the dataset. A scientist consulting herself in the development of a theory seems like malpractice. As Tomasello (2008, xiii) puts it:

> The data that are actually used toward this end in Generative Grammar analyses are almost always disembodied sentences that analysts have made up ad hoc, ... rather than utterances produced by real people in real discourse situations.

But this is too quick. The issue is quite a complex one. For example, consider the concept of observation again. In order to see just how theory-laden it is, we only have to appreciate the distinction between acceptability and grammatical judgements. Linguists distinguish acceptability of sentence or construction, which can accommodate various divergences from grammaticality, and a grammaticality judgement, which maps onto the rules of the internalised grammar.

As a clear example, L2 speakers often thwart the grammatical conventions of the language they are learning but still manage to generate intelligible or 'acceptable' utterances. However, as Scholz et al. (2022) note, there are several different ideas embedded in the use of the term 'linguistic intuitions', which include 'the judgments of (i) linguists with a stake in what the evidence shows; (ii) linguists with experience in syntactic theory but no stake in the issue at hand; (iii) non-linguist native speakers who have been tutored in how to provide the kinds of judgments the linguist is interested in; and (iv) linguistically naïve native speakers'. Obviously, (i) would be problematic from a justification standpoint. The problem with (iv) and Tomasello's 'utterances produced by real people in real discourse situations' is that they can fail to track linguistic possibility as well as fail to reflect ideal competence (given that they are error prone). Many grammatical constructions (read: licensed by the rules) either show up very rarely in corpora or sometimes not at all.

Hintikka (1999) makes the link between the reliance on speaker and especially theorist intuitions as the primary tool of theory-building in linguistics and philosophy. He accuses philosophers of being unduly influenced by the Chomskyan revolution in linguistics. For Devitt (2006), the idea is associated with a sort of Cartesianism about linguistic knowledge or privileged access to an internal state of knowledge, explicitly inaccessible. Linguistic intuitions, on this view, reflect the internal generative grammar or I-language of individual language users ('the voice of competence') and can therefore be incorporated as data for linguists. Moreover, linguists themselves can provide the data needed for their theories by introspection. Devitt's take on why linguists' intuitions are indeed relevant, *contra* the voice of competence view, comes down to a matter of expertise. A linguist surveying constructions produced by speakers is like a palaeontologist scanning a scene of ancient ruins. The palaeontologist knows what to look for and where to look for it. Similarly, the linguist, given their extensive training, is a better judge of the relevance of the data. This does not, however, mean that the linguist is or should be the sole source of such data. In fact, Marantz (2005) argues that the role of introspective data has been mischaracterised and the judgement of linguists' stand only as 'proxies' or metadata aimed at representing and not reporting. So, the judgements of linguists' merely indicate the need for further corpora-based or distributional investigation for later confirmation. On this score, studies like Sprouse and Almeida (2012, 609) corroborate this picture with their findings that 'the maximum discrepancy between traditional methods and formal experimental methods is 2%', based on textbook data points and a study involving 440 naive participants. This would lend credence to the proxy hypothesis.[20]

---

[20] For comprehensive coverage of this issue, see Schindler, Drożdżowicz, & Brøcker, (2020).

In the philosophy of linguistics, Quine (1972) directly challenges Chomsky's picture of the evidence for mental grammars. He suggests we consider what differentiates between two weakly equivalent grammars or 'behaviourally equivalent' in his parlance. These two grammars will generate the same sets of sentences. The problem is that since they are empirically adequate or generate the same sentences, there is no way of deciding which grammar is the correct description of the target or the actual grammar realised in the brain of the cogniser. Quine favoured behavioural data, while Chomskyans allow for further possible kinds of evidence (like solicited intuitions or introspection). K. Johnson (2015) applies invariance considerations from physics to mount a response to Quine's challenge. He suggests that Merge might be a tool for deciding which aspects of grammars are artefacts and which have empirical significance. Thus, the idea of 'notational variants' (or distinct grammars which are descriptively equivalent) becomes important. Nefdt and Baggio (2024a) pick this topic up and delve into the issue of the theoretical significance of the equivalent (to phrase structure) formalism of dependency grammar, often used in computational and neurolinguistic settings.

The philosophy of linguistics has investigated the question of justification in science from various angles. Work on what the evidence is, where it can be found, what it reflects when it is found, and how it relates to theory have all been thoroughly philosophically scrutinised. What's more is that this scrutiny has actually led to improved standards and more inclusion of alternative evidence-gathering methods. This change can be witnessed in a casual perusal of most theoretical linguistics journals today.

### 3.2.4 The Priority of Explanation

Formal linguistics, and the overarching cognitive revolution, emerged like many other disciplinary pursuits from the embers of logical positivism. We can see this with the general opposition to prediction as a value over or equal to explanation in linguistics (Nefdt, 2024). For the positivists, prediction was king. Theology and abstract metaphysics were rich with explanations of phenomena. But they were impoverished with relation to their predictive capabilities. Strevens (2020) considers this to be related to the empirical import of science itself, ushered in by Newton.

With Hempel and Oppenheim (1948), positivists, refashioned as logical empiricists, held prediction and explanation to be two sides of the same logical coin (cf the symmetry thesis). To see how this is the case, we need to briefly discuss the covering law or deductive-nomological model of explanation. The idea is that scientific explanations come in packages of deductively valid arguments with sound premises, in which at least one natural or general law is included. As French (2016, 55) puts it, 'we begin with the relevant laws that we explain in our theory or hypothesis and then we *deduce* from those laws a statement describing the relevant feature of the phenomenon'. Under the deductive-nomological model, prediction is explanation at a different time in the scientific process. Take the Complex NP constraint in generative grammar as an example. This rule or law states that certain types of movement or extraction from within complex noun phrase (one that includes a subordinate clause) are strictly prohibited. This law predicts the existence of 'syntactic islands' or units from which movement and extraction are impossible (Ross, 1967). The constraint can be used to explain why relative clauses work the way they do but also predict which kinds of constructions are blocked.

There are well-known problems for the deductive-nomological model, such as the fact that deduction is unidirectional, whereas explanation is not. In other words, you can deduce certain things from a set of factors in various orders that belie causal structure. For example, click sounds in Nguni languages like isiXhosa are produced when obstruents (like a fricative or plosive) collect small pockets of air and then are released in order to create loud consonants. It would be unhelpful to deduce that clicks explain the existence of obstruents (speech sounds that obstruct airflow) as this is just one (rare) case of the phenomenon. The deductive-nomological account misses the causal arrow.[21]

In any case, as Douglas (2009) argues, prediction fell out of favour due to an overcorrection in the wake of the death of positivism. Theories of explanation flourished, while theories of prediction languished. Without explicitly endorsing the symmetry thesis, Douglas motivates the need to return to prediction as a necessary tool for modelling explanation. Moreover, she goes further to claim that without an adequate account of prediction, theories of scientific explanation face serious challenges such as overconfidence or questionable connections to understanding.

> The relationship between explanation and prediction is a tight, functional one: explanations provide the cognitive path to predictions, which then serve to test and refine the explanations. (Douglas, 2009, 454)

Nefdt (2024) makes the case that formal linguistics similarly rejected computational approaches due to its overcorrection of behaviourism (Chomsky, 1957). But despite the public stance, prediction is never far from the kinds of

---

[21] This is also why generativists have argued that performance is not a good guide to competence, because the rules of competence might cause some aspects of performance but various other mechanisms intervene before the output is generated. See Dupre (2021) for a clear exposition of this argument.

theories that linguists posit. For example, Egré (2015) argues that the notion of prediction is equally applicable in linguistic explanations as it is in other empirical scientific contexts. He shows this by considering Halle's analysis of plural formation, that 'any nontrivial descriptive generalization will be predictive, provided it is testable on cases not initially considered in the inductive basis used to make the generalization' (Egré, 2015, 455). Syntactic rules operate similarly. If you state a rule, then it should predict the form that future cases will take. It's already an inductive claim, as Egré notes.

### 3.2.5 Models and Linguistic Reality

This brings us to another aspect of contemporary philosophy of science that seems particularly relevant to explanation, namely the use of scientific models. Models are surrogate systems that indirectly represent features of the target systems (Giere, 1988; Godfrey-Smith, 2006; Hughes, 1997; Weisberg, 2013). In so far as they represent those features within a more tractable, controlled space, they allow for certain kinds of predictions. During the Covid-19 pandemic, models were a major source of prediction and containment (Nefdt, 2023b). Models can be anything from sets of mathematical equations to physical materials (like the model of the double helix). As early as Chomsky (1957, 5), their worth was known in linguistics:

> Precisely constructed models for linguistic structure can play an important role, both positive and negative, in the process of discovery itself. By pushing a precise but inadequate formulation to an unacceptable conclusion, we can often expose the exact source of this inadequacy and, consequently, gain a deeper understanding of the linguistic data. More positively, a formalised theory may automatically provide solutions for many problems other than those for which it was explicitly designed.

Here, Chomsky highlights a number of themes that have received treatment only relatively recently in the philosophy of science. Firstly, for models to be useful, they need not veridicially represent their targets. Formalisation is a method for precisifying concepts and making assumptions explicit. In a classic treatment Suppes (1968) details the number of ways in which formalising aids scientific endeavour and Nefdt and Kac (2025) apply this line of reasoning to early formal linguistics. Some of these ways include generality, explicitness, standardisation, and objectivity. Generality and objectivity seem especially germane to linguistics. By the former, Suppes means avoidance of distractors, hence minimalist modelling. By the latter, he seems to be claiming that formalisation allows for a platform from which to judge certain intractable debates in science, such as the debates between behaviourist stimulus-response theorists and cognitive theorists. Müller (2018) adds computer programmability to the list of virtues of formalisation. Generative grammar was the doyen of computationalism in the early cognitive revolution. The good old fashioned AI (GOFAI) tradition drew strength from its symbolic successes towards implementation goals within machine learning and natural language processing. Without highly formalised models, this transition would be impossible to get off the ground (notwithstanding the other issues with GOFAI, see Boden (2016)).

Models have been considered mediators between theory and observation (or experimentation) (Morgan & Morrison, 1999). Knuuttila and Merz (2009) go a step further to argue that models qua 'erotetic devices' are designed not only to answer certain theoretical questions or test hypotheses but also to generate questions. Some theorists, especially within formal semantics, have explored the idea of semantics as model-based science explicitly (Jackson, 2020; Nefdt, 2020a; Yalcin, 2018). Semantics traditionally distinguishes between an object language and a metalanguage. This can be represented by the disquotational schema 'snow is white iff snow is white', where the biconditional provides the 'semantics' of the utterance on the left. The metalanguage specifies either truth conditions or other model-theoretic tools for interpreting expressions of the object language in a compositional manner. Yalcin insists that the right-hand side of the biconditional is *not* a translation of the left-hand side, from linguistic items to extra-linguistic reality. He states:

> The metalanguage of semantics, I am suggesting, is language for articulating features of the theorist's model. The interpretation function given in such a model associates pieces of language with semantic features, the latter usually modeled via certain elements of model structures in the logician's sense. (Yalcin, 2018, 355)

Similar accounts can be given in pragmatics. Stalnaker (2002)'s explanation of the dynamics of conversation in terms functions on an ever-updating common-ground defined as sets of possible worlds is clearly a model. In fact, when pressed on the metaphysics of possible worlds in this model, Stalnaker (1976) assumed a much more moderate nominalist view (in contrast to the extreme modal realism of Lewis (1986)). Modelling is pervasive in linguistics as it is in many other special and natural sciences. Most models, perhaps with the exception of some very abstract thought experiments, don't just explain they also predict. Something as strong as the symmetry thesis might be unlikely but an exclusive explanatory platform is equally so.

Analytic philosophy, following on from logic, also tends towards a preference for explanatory models and methods over predictive ones. This could be a feature of the armchair modality or the first principles approach in general. But the avowed naturalism of linguistics should disabuse linguists of this methodological restriction. Chomsky (2000) criticises the philosophy of language for its 'methodological dualism' in treating language (and mind) as special in some a priori sense, immune to the empirical standards of the natural sciences. Similarly, the recent naturalistic movement in philosophy embraces experimental methods and, in cases like Machery (2017), pushes for a picture of tractable, testable models being extracted from otherwise abstruse and irreconcilable philosophical doctrines.

## 3.3 Realism versus Antirealism

A book about the philosophy of science that does not talk about the scientific realism would ... probably be fine and cover other issues. But that doesn't mean that scientific realism hasn't commanded significant amounts of conceptual space in the field. It would not be an exaggeration to say that it underlies most of the issues we have encountered so far. Does the predictive successes of science mandate some ontologically committing stance on the objects it posits? What are scientific explanations if not descriptions of some (human) mind independent reality that go beyond our instruments of discovery? Does paradigm shift and theory change affect the veracity of current theories? All of these topics are important and lucky enough to feature prominently in countless textbooks and monographs. So we will restrict ourselves to the ways in which this larger issue manifests in the philosophy of linguistics.

Before we embark on this brief sojourn, some specifics on what exactly scientific realism amounts to. Devitt (2005), our part-time philosopher of linguistics, is useful here. First he distinguishes between 'common-sense' realism and scientific realism. The former seems committed to the existence of observable entities of both common sense and science. But Devitt's view on scientific realism is that it is only committed to the existence of the unobservables posited by our *mature* theories. This latter move is meant to soften the blow of pessimistic meta-induction (PMI) to a certain extent.[22]

We see this form of scientific realism quite ostensibly in the philosophy of linguistics. Theorists often state commitment to movement, hierarchical structure, and Merge but deny the significance of surface forms. Chomsky (2000) is clear that our intuitions about language (and mind) are bad indications of the real linguistic value. In fact, the very distinction between E-languages, which are amorphous public entities, and I-languages, the unobservable structure of the mature state of the language faculty, is illuminating. The idea that latter are meant to be the true objects of scientific linguistics rests on the idea that

---

[22] Indeed, in its original form, the PMI seems too strong. It simply is not the case that all of our fallen scientific theories were equal in quality nor that they have all failed to capture elements of reality entirely.

the unobservables reflect a deeper reality. Rey (2020) is an interesting case. He supports mainstream generative linguistic theory but disagrees on the ontology of 'standard linguistic entities' like noun phrases and phonemes. He considers such objects to be like Kanizsa triangles or optical illusions that explain intentional contents but do not map onto anything in the world. He calls these entities 'intentional inexistents'. The position is intriguing. It eschews ontological commitment to the objects of linguistic theory while maintaining a realist stance towards linguistic theory in general.

What about languages themselves? Are they real? Well, this is beautifully complicated in the philosophy of linguistics. For mentalists, language is both real and mind-*dependent* (there would be no languages without human minds). But public languages, or E-languages, are not *scientifically* real even if they might be thought of as *commonsensically* real. Just to add even more flavour. Chomskyans seem to admit their own theoretical posits are proxies for future neurological events and processes. Chomsky states, of Merge and other posits, that 'if we want a productive theory-constructive [effort], we're going to have to relax our stringent criteria and accept things that we know don't make any sense, and hope that some day somebody will make some sense out of them' (Chomsky & McGilvray, 2012, 91). Behme (2014), in her very critical review of the book from which the quote comes, asks how it is possible to defend a view that accepts nonsensical things. But the history of quantum mechanics provides ample examples of reluctant realism and proclaimed future resolutions to ontological puzzles.[23] We will return to some of these prospects in Section 6.

The standard arguments against scientific realism have unique instantiations in the philosophy of linguistics. And more intriguingly, the most common argument for it has limited scope. As previously mentioned, Lappin, Levine, and Johnson (2000) argue that generative linguistics itself underwent radical theory change from the early theories (the Standard, Extended, and Government and Binding) to minimalism. Generativists insist that minimalism just marks a shift from descriptive adequacy to explanatory (or beyond explanatory) adequacy. This means that the theory moved from concerns about describing the acquisition data, a task that could marshal significant innate resources, to one that aimed to explain the evolution of language, a task that requires a vast reduction of such resources. This move also confronts the issue of the underdetermination of theory by data. What allegedly makes minimalism a 'better' theory than earlier ones (or non-generative rival

---

[23] See Becker (2018) for a fascinating history of quantum mechanics and its 'shut up and calculate' mantra.

accounts) is its adherence to simplicity and optimality as biologically constraining features. Chomsky (2005) outlines what he calls third factor principles that shape the language faculty. Economy principles, efficient computation, genetic endowment, and a host of others all form part of these language-independent principles that nonetheless have shaped language evolution. Thus, two theories that are descriptively equivalent (or equally empirically adequate in the parlance of the philosophy of science) can be separated by means of third factor considerations. It seems both the PMI and the underdetermination of theory by evidence have special significance in the philosophy of linguistics.

One of the strongest arguments put in favour of realism is the 'no-miracles' argument (Putnam, 1975). The argument states that realism is the only game in town, philosophically speaking. If we want to explain the successes of science, anti-realist views are simply at a loss, rendering the many and major achievements serendipitous or even miraculous. The argument, however, turns on more than just explanatory success. Science has found vast technological applications from space exploration to smart phone technology. These successes required significant predictive and experimental advancements. This is where theoretical linguistics falters to some extent. Despite the initial promise, language technology, psycho- and neurolinguistics have not greatly benefitted from formal theoretical models. A well-known anecdote attributed to Frederick Jelinek, which made the rounds in computer science for decades, is that 'Every time I fire a linguist, the performance of the speech recognizer goes up' (Jelinek, 1988). Returning to miracles, the good part is that the argument does not entail only one kind of realism about science. Theoretical models might still have a shot at a different sort of realism.

Nefdt (2021, 2023a) offers a structural realist alternative to addressing both the PMI and the underdetermination issue (under the guise of multiple theories problem). Following French (2014), Ladyman (1998), and Ladyman & Ross, 2007), he suggests that structural realism in linguistics too presents a 'best of both worlds' strategy (Worrall, 1989). For generative linguists, it provides much-needed structural continuity (of which there is much to find), and for the connection with other linguistic theories, unification is offered as a model of explanation (Kitcher, 1989). On this view, a language is like a real pattern, in the sense of Dennett (1991), and a grammar a special kind of compression algorithm.

Constructive empiricism has not been explicitly applied to linguistics as far as I can tell. This is strange given its dominant status as an anti-realist alternative and its relevance to many views about language. It's chief proponent is van Fraassen (1980), who laid the groundwork for a complex account of science. The mantra for constructive empiricism is that 'science gets the observables right'. In other words, empirical adequacy is the goal of science. But along the way, the view makes a few interesting moves. It adds a dose of scepticism to the unobservables (dovetailing with Rey (2020)), a literalism about the language of science, and a reduction of theoretical terms to lists of observation statements. Many linguistic theories, so-called surface grammars, adopt a similar stance. American structuralists and probabilistic grammarians aim to get the observables right and are reticent about what underlies them (at least what entities). Dependency grammars posit no underlying structures or invisible movement operations (Nefdt & Baggio, 2024b). They operate purely on the level of observable words and morphemes. In the literature, this is called a lexicalised grammar. There are many such frameworks to choose from.

The key to constructive empiricism is the rejection of the criterion of truth (or even approximate truth) for our scientific theories. In its place is a modal concept of what the world *could* be like (without ever telling us exactly which world we live in). This strategy obviates both the PMI and underdetermination to different degrees. Since scientific progress is just the accumulation of observable regularities or results, the unobservables can change as they like. We also don't have to choose between rival theories which are empirically equivalent unless we have some pragmatic grounds for doing so. In terms of explanation, constructive empiricism was partly responsible for the boom in work on scientific models and their role in scientific explanation and prediction.

For constructive empiricism to help the philosophy of linguistics, we might need a clearer notion of the observable—unobservable distinction. But here, the philosophy of linguistics might be able to provide some assistance itself. Notice how many linguistic concepts are buried within this theory. Besides the pragmatics of theory choice and the semantic literalness of scientific discourse, linguists have been labouring over the observable—unobservable distinction for decades in the form of the competence—performance distinction (assumed in many if not most theories to varying degrees). Reflections on the relationship between ideal competence and actual performance in linguistic theory could shed light on how to draw the distinction elsewhere.

There is more to unpack here. One further curiosity is that platonists, mentalists, and nominalists in the philosophy of linguistics all make use of the term 'realist' to characterise their views. So the realism is really about what to be realist about, sets, mental grammars, or splotches of ink on paper. We will have to leave matters here for now as the metaphysics of language is beyond the current scope. In the next section, we explore a case study in the philosophy of linguistics, one that speaks directly to intricate issues in the philosophy of science (and mathematics) but rarely finds representation in those spaces.

## 4 Case Study I: Infinite Generalisation in Linguistics

This case study is meant to exemplify an issue in the philosophy of science concerning the use of powerful formal apparatus in describing and explaining natural phenomenon. Wigner (1960) initially asked a series of questions about why (and how) abstract mathematics is so 'unreasonably' effective in the physical sciences. His two central explananda were (1) 'that the enormous usefulness of mathematics in the natural sciences is something bordering on the mysterious and that there is no rational explanation for it' and (2) 'it is just this uncanny usefulness of mathematical concepts that raises the question of the uniqueness of our physical theories'. Recently, philosophers of science have taken these puzzles as a challenge. Drawing from structuralism in the philosophy of mathematics, Pincock (2007) explores the extent to which a mapping analogy is explanatorily fruitful. Bueno and Colyvan (2011a), Bueno and French (2018) produce an inferential framework for explaining the application of mathematics which involves three stages: immersion or the initial mapping, the derivation step which is when theorists draw mathematical conclusions in the model and finally the interpretation step in which the previous conclusions are interpreted in the empirical or assumed structure. Nefdt (2020a) attempts an application of this inferential framework to formal semantics in which lambda theory, set theory, lattice theory, and a number of other tools have been harnessed in its analysis (see Landman, 1991; Partee, ter Meulen, & Wall, 1990). In this case study, we focus on one powerful tool, set theory, and one powerful concept, namely infinity or infinite generalisation, as they are found in theoretical linguistics.

Despite the various expositions and ruminations on the concept of linguistic infinity, few philosophers outside of the field seem to appreciate its contribution to the larger scientific debates on infinite generalisation. In an otherwise excellent article in defence of the notion of potential infinity in the philosophy of mathematics, Linnebo and Shapiro (2017) devote a single footnote to linguistics. They state:

> The notion of potential infinity is still with us, perhaps in a more subtle form. It is now a commonly held view in linguistics that languages are infinite. Noam Chomsky, for example, once wrote that a grammar projects from a finite corpus to a 'set (presumably infinite) of grammatical sentences' (p. 15). It is, we think, more natural to think of a language as potentially infinite. (Linnebo & Shapiro, 2017, 187)

The question of infinity is not just a mathematical puzzle, at the very heart of set theory but also a scientific quandary about the nature of idealisation. Consider phase transitions used to explain the breakdown of electromagnetic gauge invariance in physics. Phase transitions involve a thermodynamic limit or, 'in other words, we need to assume that a system contains infinite particles in order to explain, understand, and make predictions about the behaviour of a real, finite system' (Morrison, 2015, 27). In population genetics, the investigation of finite real populations results in models of infinite populations and their properties. Even engineering employs infinite idealisations in its structural analysis (Bianchi, 2019).[24]

Infinity and infinite generalisation form both an essential concept and an indispensable tool in mathematics and science. In order to appreciate how linguistics has interrogated the concept and used the tool, we will first briefly start with the basic mathematics of infinity.

## 4.1 What Is Infinity?

There are many excellent introductory texts on infinity in mathematics.[25] The notion of the infinite has been prominent in philosophy, theology, and scientific thought for centuries. However, it wasn't until the work of Georg Cantor in the late 1800s that mathematics truly captured the idea precisely. One particularly central insight or discovery was that there are different sizes (or cardinalities) of infinity. Let us start by distinguishing finitude, countable infinitude, and uncountable infinitude.

Consider the famous thought experiment of Hilbert's hotel.[26] If there was a hotel filled to capacity with *finitely* many rooms, any newly arriving guests would have to be turned away. However, if the hotel was *infinite* in that it had rooms for every natural number (starting from 0), such that guest 0 occupies room 0, guest 1 occupies room 1, and so on, it could be full and still admit new guests. How is this possible? Well, imagine a new guest named Georg shows up and asks for a room. The hotel manager would merely have to ask every current guest to move up one room from their present location. This procedure will free up room 0 (as the former occupant of room 0 would now be in room 1 and so on). This only works because the axioms of Peano arithmetic guarantee us that every number has a unique successor, so there is no largest natural number. In fact, no matter how large the finite magnitude of guests that arrive at the hotel, our trusty hotel manager can accommodate them by merely relocating guest $n$ to room $n +$ that number. If an infinite number of guests arrive, a similar procedure can be implemented. Simply move guest $n$ to room $2n$, thus freeing up

---

[24] For a special issue on infinite idealisations in science, see Fletcher et al. (2019). The issue does include an article on infinite generalisation in linguistic by the present author, Nefdt (2019a).

[25] See Oppy (2006) and Easwaran et al. (2023) for an entire disquisition on the concept of infinity and Rayo (2019) for an accessible introduction. For more general set theory, see Jech (2005).

[26] It hasn't always been clear if Hilbert himself was actually responsible for the idea. However, see Kragh (2014) for a history of the hotel that finds Hilbert at its source.

an infinite number of rooms for occupancy. The question then looms, are there any magnitudes of guests that our hotel cannot accommodate according to this procedure? To answer this question, we need to introduce some terminology.

The *cardinality* of a set (represented as $|A|$ for set $A$) tells us how many members the set contains. In set theory, the standard mechanism for determining the relative size or cardinality of sets is *bijection*. This is an equivalence relation (that is a reflexive, symmetric, and transitive relation) that ensures that between two sets $A$ and $B$, there is a function that maps each member of $A$ to a different member of $B$, and no member of $B$ is left behind in the pairing. In finite sets, this means that two sets have the same of cardinality iff there is a bijection between them.

A number of infinite sets can be shown to have the same cardinality by establishing bijections between their members. Consider the set of natural numbers $\mathbb{N}$ and the set of integers $\mathbb{Z}$ (containing the natural numbers and their additive inverses). The procedure for showing these sets to be of the same size is relatively straightforward. One way of doing it is by assigning all the even numbers from $\mathbb{N}$ to all the positive integers and the odd numbers to the negative ones. Showing that there is a bijection from $\mathbb{N}$ to $\mathbb{Q}$ is also relatively simple (although it involves ignoring some identical fractions with different labels like 1/2 and 2/4). This brings us to the definition of a countable infinity, that is, any set that can be shown to have a bijection with $\mathbb{N}$.

Returning to our question of whether there are any infinite sets of guests that would have to be turned away from Hilbert's Hotel, a result known as Cantor's theorem sheds the necessary light. Basically, Cantor's theorem states that sets have more subsets than members or 'for any set $A$, $|A| < |\mathcal{P}(A)|$' (Rayo, 2019, 11).[27] This results in an iterative infinity of infinite cardinalities. More importantly for our purposes, it distinguishes the real numbers $\mathbb{R}$ from $\mathbb{N}$. In fact, it turns out that $|\mathbb{R}| = |\mathcal{P}(\mathbb{N})|$. That means that the set of real numbers is strictly larger than that of the natural numbers, it is *uncountable*. In other words, if a busload of guests show up with the cardinality of real numbers, our hotel manager will be in real trouble.

An important fact to note is that once we move past finite magnitudes, intuition begins to fail us. Specifically, our intuitions on relative sizes of sets seem to imply what Rayo calls the 'proper subset principle': if $A$ is a proper subset of $B$, then $B$ is larger (has more members) than $A$. When we reach infinite sets, this principle contradicts the bijection principle, that is, that two sets are the same size if there is a bijection between them. For instance, the set of natural numbers is a proper subset of the set of rational numbers ($\mathbb{N} \subset \mathbb{Q}$) but, according to

---

[27] $A$'s powerset ($\mathcal{P}(A)$) is just the set of all the subsets of $A$.

Cantor's conception of sameness of size, is of equal cardinality (or even more starkly, the even numbers are a proper subset of the natural numbers yet have the same cardinality). The set of real numbers can be constructed by defining Dedekind cuts on the rational numbers. But it isn't the fact that the set of natural numbers (or rationals) is contained in the set of reals which makes the latter uncountable but rather what seals its infinite fate is that lack of a bijection.

## 4.2 Linguistic Infinity

When generative linguists talk about infinity, they often use the term 'discrete infinity' to mean a countable infinity. Recently, Chomsky et al. (2023, 1) have once again placed the concept at the centre of the field:

> The infinity of sentences is the basis for this ethical principle; there are enough sentences to go around for each of us to use only our own. It's also the starting point of Generative Grammar, and of our discussion of Merge: a driving goal of the modern study of language is to determine and explain this property of discrete infinity.

Chomsky himself has repeatedly stated that infinity marks a unique departure from other biological phenomena and systems in nature.[28]

> Language is, at its core, a system that is both digital and infinite. To my knowledge, there is no other biological system with these properties. (Chomsky, 1991b, 50)

> Some basic properties of language are unusual among biological systems, notably the property of discrete infinity. (Chomsky, 1995b, 154)

Let us call this the *linguistic infinity postulate* (LIP). There are a number of popular avenues for establishing LIP (although it is often just stated without argument). From the literature, there is one usual strategy for establishing this result. It attempts a proof-theoretic reconstruction of an insight, attributed to Willem von Humboldt, that language is 'infinite use of finite means'. This strategy allegedly produces the requisite analogy with the natural numbers.

The literature on recursion in linguistics is complicated. Specifically, it is unclear at which level (or even which object) the property is meant to be interpreted. I cannot embark on a mission to disentangle the many vines of this particularly thorny issue here (see Tomalin (2006) and Lobina (2017) for some perspective). Basically, the mathematical definition involves some property of self-reference, usually in two steps: one that specifies the condition of termination of the recursion (or the base case) and the recursive step which reduces

---

[28] 'Digital' here means 'discrete' or composed of atomic units (as opposed to continuous as in the real numbers). We'll ignore the claim of biological isolation for our current purposes, see Nefdt (2023b) for more on this issue.

all other cases to the base. For example, addition in Peano arithmetic can be defined recursively as $x + 0 = x, x + Sy = S(x + y)$ (where '$S$' denotes the 'successor of' one place relation). Similarly, the rewrite rules for a context-free grammar are recursive: $S \to ab$ or $S \to aSb$ (where 'S' is now a unique start symbol or category and 'a' and 'b' are terminal elements like words).

This brings us to the first argument in favour of discrete infinity, that is, that generative grammars output infinite expressions or grammatical structures. If a grammar $G$ is a set of finite rules for defining stringsets (or sequences of symbols), then a language is the set of all the strings that $G$ generates. In post-production system style, '$G$ will be said to generate a string $w$ consisting of symbols from $\Sigma$ [the alphabet] if and only if it is possible to start with $S$ and produce $w$ through some finite sequence of rule applications' (Jäger & Rogers, 1957, 2012). What does this mean for LIP? Well, it comes down to what is known as the 'membership problem'. Generative grammars have sharp boundaries. Some string $w$ is either generated (or derived) by the rules of $G$ or it isn't. It is either in the language or it is not. There is no in between. This means that the language that the grammar generates 'has an exact cardinality: it contains either some finite number of objects or a countable infinity of them' (Pullum, 2013, 496). Furthermore, the reason for opting for a countable infinity (i.e. LIP) allegedly comes from the data of natural language itself.

In order to establish a connection between the set of natural language expressions and the set of natural numbers, we need a few things. Firstly, we need an analogy with the successor function in Peano arithmetic. Remember, the successor function applies to an object in the set and produces another object in that same set endlessly. To mount an analogy, we would need a similar kind of operation in language. The usual candidates come from syntax.

The idea is something like: English syntax is 'closed under adverbial modification' (or coordination or subordination, etc.). In other words, if 'Noam is brilliant' is in the set of English sentences or, if you prefer, generated by the rules of English, then so is 'Noam is very brilliant', 'Noam is very very brilliant' so on ad infinitum. The same strategy can be employed for other common examples involving subordination, centre embedding, conjunction, and so on. Pullum and Scholz (2010) call this the 'no maximal length claim' (NML) or the claim that there is no longest English sentence (i.e. you can always create a longer one). The problem they point out is that to establish this analogy, one would have to assume both that every English expression has a successor, and that this function is injective somehow. But doing so would essentially be begging the question.

It should be noted that Chomsky (2008) does attempt to make a direct argument concerning the Merge operation of Minimalism and arithmetic. He outlines the following procedure for mimicking the successor function within linguistics.

> Suppose that a language has the simplest possible lexicon: just one LI [lexical item], call it 'one'. Application of Merge to that LI yields {*one*}, call it 'two'. Application of Merge to {*one*} yields {{*one*}}, call it 'three'. Etc. In effect, Merge applied in this manner yields the successor function. It is straightforward to define addition in terms of Merge(X, Y), and in familiar ways, the rest of arithmetic (138).

The idea is that Merge applied over a lexicon with a single entry results in the successor function. Mukerji (2022) and Hinzen and Uriagereka (2006) too make the connection between arithmetic and language. However, the problem is that Merge, or the set-theoretic operation of taking two objects and creating a new labelled unit with the two objects as (unordered) members, has nothing to do with language in and of itself. Making the argument that natural language syntax can be characterised with this simple operation is a separate task (one that minimalists have taken up for twenty years; see C. Collins and Stabler (2016) for a clear formalisation). But merely showing that you can model the successor function with a recursive set-theoretic operation does not settle the matter.

This brings us to an important assumption about the nature of generative grammars imbued with recursive rules, namely that they necessarily yield or entail a discrete infinity of linguistic outputs. Again, Pullum and Scholz (2010) show that this entailment doesn't hold by means of a toy example using a context-sensitive grammar. The grammar contains a recursive rule ($VP \rightarrow VP\ VP$) which is only capable of a single application given the specified lexicon, but nevertheless remains non-trivially recursive. Thus, the recursive element of the grammar does not ensure infinite output (even if it can create infinite structures).

More recently, Huybregts (2019), a firm adherent of LIP, insists that infinite cardinality is the default or null hypothesis of formal grammars. His target is evolutionary accounts of Merge as a step-wise or gradualist approach to infinite language along a cultural or evolutionary path (starting with some finite magnitude). He reiterates NML as a datum. What is interesting is where he locates the infinitude. He states '[w]e should be careful, however, not to confuse the infinite productivity of the generative procedure (a function in intension) with the cardinality of the decidable set of structures it generates (the well-formed formulas defined in extension)' (Huybregts, 2019, 3). Appreciating this caveat apparently allows for explanations of languages with finite output or no overt recursive constructions, without giving up on LIP. This relates back to our discussion of standards of evidence in Section 3.2. Hence, the counterexample provided by Pullum and Scholz (2010) is apparently neutralised since the grammar 'may generate an infinite language but only produce a finite subset of it' (Huybregts, 2019, 3). If this claim holds, it is because linguistics is about structures not stringsets. Everaert et al. (2015) make this position explicit. They argue that it is a mischaracterisation to associate linguistic theory (and natural languages) with the generation of stringsets. The computations of the mind can only 'see' hierarchically structured, recursive phrases. This is enough to establish an infinity without postulating actual sets of collections of sentences. The early distinction between weak, involving the generation of strings, and strong generative capacity, involving the additional association with structures (like trees), might be helpful here.[29]

Not only does Chomsky provide us with arguments as to the adequacy of different formal language families but in his landmark Chomsky (1959a), he takes one step further to define a nested hierarchy of computational complexity showing the relationships between these formalisms. At the bottom or innermost ring are the regular languages (Type-3). They are easy to parse but not powerful enough to capture various patterns of natural language. Context-free languages do better but are more complex as a result (and contain the regular languages as a proper subset) (Type-2). Context-sensitive languages can capture patterns that are missing from CFGs (such as $a^n b^n c^n$) but do so at the cost of computational tractability (Type-1). Lastly, the class of languages that comprise all of the other families and constitute all the computably enumerable languages are Type-0 grammars. A useful way of understanding this level is by appreciating their associated accepting automata. In this case, the automata they map onto are unrestricted turing machines. Chomsky claims that unrestricted turing machines offer us nothing of interest in terms of linguistics. 'We learn nothing about a natural language from the fact that its sentences can be effectively displayed, i.e. that they constitute a recursively enumerable set' (Chomsky, 1959a, 138). Importantly for him, for the study of formal languages to be relevant, it needs to not only describe stringsets ('weak generative capacity') but also provide structural information ('strong generative capacity'). Type 0 grammars are unrestricted and thus do not perform that latter task. Chomsky thinks that the question of where natural language lies in the hierarchy of formal languages is a significant one.[30] The reason he provides is interesting. He insists that an adequate theory must serve as a basis for an account of how speakers understand

---

[29] See Chomsky (1963) for an initial discussion and P. Miller (1999) for a model-theoretic alternative characterisation.

[30] Today it is assumed that natural language syntax falls somewhere between CFLs and context-sensitive languages in the so-called mildly context-sensitive class (Joshi, 1985). See

and use a particular grammar, that is, they must be empirically motivated. But this opens up the possibility that some speakers do not make use of languages with recursive structure at all. The possibility is chased up by Langendoen (2010). He describes the situation in the following way.

> For any natural language $L$, let $L^\square$ represent the finite set of expressions known to belong to L on the basis of such direct evidence. Given that $L^\square$ provides indirect evidence for genuinely recursive size-increasing operations, standard generative models project a denumerably (countably) infinite set $L^\diamond$ of 'possible' members of L. By not distinguishing the models from the language, proponents of such models conclude that $L = L^\diamond$ and so is infinite. As noted above, that conclusion may be correct, but an argument is still needed to show that the models do not overgenerate. In the absence of such an argument, all we can conclude is that $L$ lies somewhere between $L^\square$ and $L^\diamond$, and so may be either finite or infinite.

This formalisation allows for some languages to be modelled as infinite and others not (based on the evidence of the respective linguistic communities). This kind of modelling makes linguistics somewhat unique. Highly formal concepts usually resigned to a priori mathematics receive empirical justification and real-world instantiation not just application (as in the natural sciences). The philosopher of mathematics, Resnik (1982, 101), describes the situation well.

> Take the case of linguistics. Let us imagine that by using the abstractive process... a grammarian arrives at a complex structure which he calls English. Now suppose that it later turns out that the English corpus fails in significant ways to instantiate this pattern, so that many of the claims which our linguist made concerning his structure will be falsified. Derisively, linguists rename the structure *Tenglish*. Nonetheless, much of our linguist's knowledge about Tenglish *qua pattern stands*; for he has managed to describe *some* pattern and to discuss some of its properties. Similarly, I claim that we know much about Euclidean space despite its failure to be instantiated physically.

According to this view, linguists, like mathematicians, study patterns (Hellman, 1989; Nefdt, 2018, 2023a; Shapiro, 1997). These patterns are formally characterisable. But their formal properties stand in a modelling relation to the world. Careful study of the actual requirements of individual language communities (or individual cognisers) will reveal the extent to which the formal properties are instantiated in the real-world languages under study. Of course, if corpora decide the empirical details, then infinite generalisation is unlikely to be found in the natural systems in the same way that phase transitions do not

---

Jäger, G., and Rogers (2012) for more on the refinements of Chomsky Hierarchy and Graf (2022) on the linguistic fecundity of studying subregular languages.

support the existence of infinite particles in a finite universe. However, if the abstract rules of competence ultimately decide the remit of the grammar, then some sort of potential infinity might be more realisable.

## 4.3 Lessons for the Philosophy of Science

There are a number of lessons to be drawn from the case study discussed in Section 4. Some are negative. Formal linguists have been accused of possibly conflating aspects or artefacts of their formal models with elements of the target system. If generative grammars are models, then they might indeed entail infinite output. But this does not hold immediately for the natural languages that they model (ignoring the technicalities of whether it holds in general). It might just be more convenient to talk *as if* natural languages were infinite (Appiah, 2017).

On the positive side, the linguistic infinity postulate does seem to account for the data on iterative structures in competence, not to mention the phenomenology of linguistic creativity. Perhaps human language is biologically unique and the philosophy of science (and biology) should pay due attention to the possibility of a genuinely infinite structure in the human brain (and the universe). In many ways, linguistic infinity or infinite generalisation provides a rich set of arguments and evidence for the existence of infinite sets or objects in the physical world, whether or not you agree that language is infinite itself. The situation can be well characterised in terms of the concept of neutral analogy in Hesse (1963). For Hesse (1963), neutral analogies (as opposed to positive or negative) involve cases in which it is unknown whether the model and the target share certain properties. Thus, they offer a route to possibly new hypotheses. A classic example is Dirac's discovery of the positron, in which negative energy solutions were initially considered to be mere features of the mathematical model (or 'artefacts'). However, after finding physical interpretations of these solutions, he was led to revise his entire theory and predict the existence of a novel particle (Bueno & Colyvan, 2011b).

The concept of linguistic infinity might not answer the puzzle concerning the effectiveness of set theory or mathematics in general within the sciences but it does provide fertile ground for further exploration of the topic of infinite generalisations and idealisations.

## 5 Case Study II: Language Models and Linguistic Theory

The previous case study delved into the deep, foundational mathematical bedrock of linguistic theory. The applications of proof theory, set theory, and other mathematical tools drew their initial inspiration from the tectonic shifts within mathematics and science in the early twentieth century. There, I used the concept of infinity as a connection point to larger issues of idealisation and infinite generalisation in the sciences, with linguistic theory as Sherpa. In this case study, we shift focus to more engineering concerns with no less significance to the philosophy of linguistics. Recently, discoveries and advances in AI have fuelled debates on the relevance of large language models to linguistic theory (Baroni, 2022; Moro, Greco, & Cappa, 2023; Piantadosi, 2024). To what extent are innovations of technology and engineering relevant to the theory of language or to cognitive science in general? Whereas Section 4 can be seen as a case study in the use of formal, discrete mathematical formalisms in science via the philosophy of linguistics, this section can be read as a case study in the role of statistical, engineering models in science. In this way, we may gain useful insights into the nature of scientific modelling in the philosophy of science. Again, the philosophy of linguistics will be our guide.

Despite the explanatory goals of symbolic processing, inspired by the influence of computationalism in linguistics among other things, GOFAI had limited success in terms of natural language processing (NLP). The modern variant of AI responsible for significant engineering successes is based on the foundations of connectionism as well as continuous mathematics and advances in GPU hardware. Deep learning AI is an intricate methodology with part calculus, linear algebra, vector-space distributional semantics, artificial neural networks, and statistics over massive corpora. The basic architecture involves neural nets consisting of input layers fed to multiple (again, sometimes in the hundreds) hidden layers, culminating in a probability distribution or output layer (which could also be binary). Deep learning goes well beyond language models but the main task for LLMs is autoregression or next token prediction. At this task, they are exceptional. What's more surprising is that being exceptional at such a task has somehow unlocked syntactic fluency in language production.

There are so many angles to pursue on this huge topic.[31] But this is not a philosophy of AI or deep learning book. Thus, we'll limit ourselves to how this remarkable engineering achievement might affect linguistic theory, if at all.

Two camps seem firmly established. On the one side are those who not only believe that LLMs can serve as scientific theories of language, such as Baroni (2022), but also entertain the thought that they already possess enough acuity to challenge (generative) linguistic theory. In a tendentious article, Piantadosi

---

[31] For more on the history and mathematics, see Ananthaswamy (2024). For an introduction to LLMs from a philosophical perspective, see Millière and Buckner (2024).

(2024) details the various advancements of LLMs and argues that they effectively refute the strong innateness hypothesis and the poverty of stimulus arguments. He considers them profoundly superior to any and all linguistic theory: 'game changers'. His position has engendered numerous equally strong responses (Katzir, 2023; Kodner, Payne, & Heinz, 2023).

On the other side, some linguists, following Chomsky (2023), and Chomsky, Roberts, and Watumull (2023), have insisted that LLMs offer absolutely no relevant structure, information, or tools for linguistic theory. The chief argument at the moment is that LLMs fail to distinguish possible from impossible human languages. Moro, Greco, and Cappa (2023) make this case explicitly and rely in part on work done by Mitchell and Bowers (2020), showing that recurrent neural networks (without strong inductive biases) can learn number agreement in both normal language settings and within unnatural sentence structures. Kallini et al. (2024) test this claim further with more contemporary transformer-based models (GPT-2 small to be precise). They run a number of experiments on an innovatively constructed continuum of 'impossible languages' to show that certain LLMs do seem to show preference for human language solutions, when primed on English data.

Of course, one of the main grievances for the cognitive sciences is that LLMs are just too large. Human children don't see anything close to the amount of data that LLMs are trained on (Frank, 2023). Furthermore, LLMs are fed tokenised chunks of text. These tokens carve up linguistic reality at joints for which morphologists would lose their lunch.[32] Tokenisation is a compression procedure not a linguistically inspired categorisation. Humans also encounter language in different modalities beyond text. In fact, text comes later for us.

The real issue might come down to explanation. Can LLMs *qua* prediction machines offer us anything in the way of *explaining* the nature of human language? The moderate alternative between the two extreme camps seems to rest on the concept of 'models'. Cichy and Kaiser (2019) explicitly argue that artificial neural networks (of which LLMs are subsets) are scientific models. Millière (2025) takes up the mantel for LLMs in linguistic theory. He considers most of the topics we have been looking at in this Element, including the competence—performance distinction, language acquisition, and poverty of stimulus arguments. He suggests that LLMs can be relevant as models of competence, not just performance and thus shed light on these other problems too. But he is careful not to endorse stronger claims about their role as theories. In the rest of this section, we will rehearse some of the reasons for optimism as well as some cautionary tales. From the perspective of the philosophy of

---

[32] Or so the usual argument goes, but see Balashov (2022).

science, it certainly seems clear the LLMs can add to the toolkit of modern linguistic theory without compromising the theories which currently lead the field, and will likely continue to do so into the future.

## 5.1 Why They Matter

It might argued that LLMs are learners of human language. However, this does not imply that they learn languages as we do (see Dupre, 2024; Lappin, 2024). That is a stronger claim which comprises the core of most of the current debates.

Why think that LLMs are relevant to understanding human language in the first place? One of the most common arguments in the literature is that LLMs offer us a testing bed for cognitive hypotheses, especially with relation to the issue of innateness. A. Clark and Lappin (2012) claim that predessors of LLMs show us that only weak induction biases are necessary for language learning. Similar arguments have been made across the board. The idea that language is simply not learnable without strong (innate) structural biases is simply false. The standard counterarguments, however, point to the fact that LLMs are trained on significantly (in fact multiple orders of magnitude) more data than the average human child when acquiring their first language. To this, initiatives, such as *BabyLM*, which aim to train models on data more closely aligned to what we know about children's available stimulus were spawned. But, of course, unlike children, LLMs are generally trained on flattened text (or tokens). In response to this restriction, Vong et al. (2024) have attempted to develop a multi-modal training set involving innovative use of visual input (using head-mounted cameras on a young infant). The Zero speech challenge also aims to address this issue (www.zerospeech.com/).

The next reason for optimism comes from the success of LLMs on various linguistic tasks, even ones involving hidden structure such as *wh*-movement and other filler-gap dependencies (Linzen & Baroni, 2021; Wilcox et al., 2018). The fact that LLMs are able to pick up on subtle patterns in text only is quite remarkable. This is even possible in cases of relatively rare constructions such as English Preposing constructions in prepositional phrases or PiPPs such as 'happy as we were' or 'brilliant though they seemed', where the corpora do not exhibit many examples (see Potts, 2024).

Hasson, Nastase, and Goldstein (2020) turn the issue of over-parameterised models like LLMs (or more broadly artificial neural networks, ANNs) around to reflect an alleged analogy with what cognitive neuroscientists have discovered about brain processing. They argue that modern ANNs embrace complexity in a manner distinct from traditional experimental designs with straightforward interpretability. Rather than viewing this as a disadvantage for cognitive science, they leverage Big Data and over-parameterisation to argue that these models operate in similar ways to naturally evolved human brain networks. Specifically, they claim that both brain networks and ANNs are 'direct fit' models in which interpolation or an interpolation space is key. This process requires 'dense and broad sampling of the parameter space for reliable interpolation' (428). They contrast this with the standard assumptions in cognitive psychology of poverty of stimulus and limited computational resources leading to extrapolation criteria and 'ideal fit' models. Direct fit optimisation in brains and ANNs, in contrast, do not extrapolate from external data, but interpolate from high-dimensional dense sampling. Interpolation involves local computations which usually produce weaker generalisations. However, given enough data, that is, large training sets or Big Data, interpolation can produce the same level of predictive power as extrapolation. Furthermore, and in line with our evolutionary discussion, '[s]imilar to natural selection, the family of models to which both ANNs and BNNs [brain neural networks] belong optimizes parameters according to objective functions to blindly fit the task-relevant structure of the world, without explicitly aiming to learn its underlying generative structure' (Hasson, Nastase, & Goldstein, 2020, 425).[33] Chemla and Nefdt (2024) marshall this natural selection metaphor to caution against the idea of a general learner, pervasive in the NLP community, and suggest that both sides of the impossible languages debate are missing a step.

One core aspect of human language processing which does seem to be different in LLMs is efficiency. As we have seen in Section 2.2, prominent examples include Zipf's law, where frequency of words is efficiently associated with shorter strings, but generally dependency length minimisation (DLM) and minimum description length (MDL) are essential aspects of human linguistic cognition and analysis, respectively. In other words, in terms of the former, we tend to minimise the length between words to increase efficiency (Gibson et al., 2019). While MDL operates at the grammar or corpus level, such that optimality of analysis is defined in 'virtue of providing both the most compact representation of the data and the most compact means of extracting that compression from the original data' (Goldsmith, 2001, 2). Thus, information-theoretic concepts of compression and efficiency can explain many aspects of human language (Chater et al., 2015; Gibson et al., 2019; Nefdt, 2023a).

---

[33] Pasquiou et al. (2022) examine the extent to which LLMs fit (or predict) brain processing data such as fMRI time courses of brain activation during a listening exercise for an audio-book (a language task). Interestingly, their findings suggest no major discrepancies between different model architectures (RNNs, Transformers, and non-contextualised word embeddings). Despite this, they suggest that off-the-shelf neural language models are not fit to model brain activation patterns for language without careful training.

If human language is indeed shaped by such efficiency considerations, then LLMs should follow suit. However, research has indicated the opposite tendency. Chaabouni et al. (2019) ran experiments to show that LSTMs incorporate a preference for anti-efficient encoding rendering the most frequent expressions to the longest strings. Adding a cost function which penalises long messages moves the models towards efficiency *contra* its initial tendencies (see also Lian, Bisazza, & Verhoef, 2021).

Lappin (2024) provides a longer list of pros and cons for LLMs and linguistic theory. One of the cons that has received quite a bit of attention is metasemantic in nature. Bender and Koller (2020) argue that LLMs trained only on (textual) form cannot access semantic meaning, that is, are not semantically grounded. In addition, they lack the communicative intent required for understanding language. Their claims have been tested and challenged by using the resources of semantics and the philosophy of language. Mandelkern and Linzen (2024) argue that LLMs can access meaning indirectly via a linguistic division of labour argument from Putnam (1981). The causal externalist metasemantics they offer is in line with contemporary philosophy of language but at odds with internalists views associated with syntax. Baggio and Murphy (2024) make this case forcefully against Mandelkern and Linzen (2024). Lappin (2024) focuses more on multimodal LLMs which are largely immune to Bender and Koller's Chinese room revival.

But again, models do not need to reflect every aspect of the target system nor do they have to be made of the same stuff or even operate via the same mechanisms. The bar is relatively low for their similarity in accordance with explanatory value, depending on our scientific aims.

## 5.2 What It Means

The take home message is that the philosophy of linguistics might forever be altered by the introduction of LLMs. Despite the terminology, it is not clear that LLMs are actual models in the scientific sense. But this issue certainly needs further investigation. Without the resources of the philosophy of science, such an endeavour is more liable to cause conflation and confusion.

I think one of the clearest and most useful interpretations of the role LLMs can play in linguistic theory is that they are model organisms (Ankeny & Leonelli, 2011, 2020). This perspective is endorsed in Tuckute, Kanwisher, and Fedorenko (2024) and suggested elsewhere as well. It should even satisfy biolinguists who favour the strong interpretation of their field. The idea is that only limited comparative work was possible with our primate cousins and animals further up (or down) the phylogenetic tree. Their capacities for anything resembling the complex structures of natural language are impoverished at best. For the first time in our history, another system seems able to reproduce (or mimic) of linguistic behaviour in much of its complexity. Probing and prodding this system might not yield direct homologies with the human language system but it could tell us something interesting and useful about linguistic representation itself. We have learned much from comparative work in vision. Language until now has be more refractory given its relative uniqueness within the biological world. The analogy goes potentially deeper. Different LLMs incorporate different architectures and process language is distinct ways. Recurrent networks include a memory element and were generally the standard models for NLP before transformers with their attention heads and positional encoding proved more robust on most linguistic tasks (Vaswani et al., 2023). Studying how different models operate, instead of just moving on to newer ones based on their successes on performance benchmarks, would enrich this avenue for linguistic enlightenment further and push the model organisms perspective to its limits.

One might object that model organisms only provide useful information in so far as they are biological, organic creatures linked to us by genetics, natural selection, and evolution. Searle (1980) seemed to suggest that machines are just not made of the right kinds of stuff to be considered conscious or capable of understanding. Oddly, this biochemical chauvinism is a difficult position to maintain in principle, especially for those who prescribe to a genuinely computationalist perspective on mind and language. The LLMs are nothing if not computational, albeit in the non-classical, connectionist sense of the term. It is still true that LLMs are certainly not occupants of the phylogenetic tree of life that humans and other non-human animals both find their respective branches. Nevertheless, as mentioned in Section 5.1, Hasson, Nastase, and Goldstein (2020) and Chemla and Nefdt (2024) do find analogies with natural selection. In fact, Yax, Oudeyer, and Palminteri (2024) push this metaphor to its extreme, as they propose a framework for constructing phylogenetic trees of LLMs (primarily in terms of 'traits', that is their behaviour as to what tokens they generate and how they fare with respect to various benchmarks, rather than direct structural or historical considerations). It might be useful to note that many model organisms are often carefully bred to elicit the comparative properties we want to model, resulting in significant departures from their wild counterparts.

Despite this possible positive, one can sympathise with those who emphasise the successes of LLMs, where traditional symbolic models have failed. If natural language is indeed based on discrete rule systems somehow embedded in our neural pathways, then achievements of LLMs which do not rely on these kinds of mathematical properties are even more puzzling. But, of course, statistics and Big Data can go a long way towards approximating intelligence without ever reaching it. The issues are thorny. Is this a case of the detachment of prediction from explanation in the philosophy of science (discussed in Section 3.2)? Or are LLMs telling us something important about ourselves? Here, the honed work of philosophers of linguistics, steeped in the philosophy of science, can help clarify confusion and chart new paths forward. Or at least, one hopes it can.

## 6 The Philosophy of Linguistic Subfields and Future Prospects

In lieu of a conclusion, I will offer a list of honourable mentions of work in the philosophy of science that covers other linguistic subfields before speculating on what the future might hold for the philosophy of linguistics.

Before I list the candidates, I should mention that my focus on the philosophy of science (or a particular interpretation thereof) has obscured some excellent work on the metaphysics of linguistic objects and the type-token distinction (Bromberger, 1989; Kaplan, 1990; Katz, 1981; Wetzel, 2009). Although these topics are separable, they often find common expression within the growing literature on the philosophy/ontology of words (see a special issue in *Synthese* by Miller and Hughes (2023))[34]. Nefdt (2019b) applies structuralism in the philosophy of mathematics to words, Mallory (2020) explores linguistic types (like word types) as action types, J. Miller (2021a, 2021b) looks at the connections between words and species from within a bundle theory perspective, J. Collins (2023b) takes a Chomskyan internalist view of words, and Stanton (2023) looks at the prosodic behaviour of certain intensifiers as compared to non-words, and reveals novel intricacies in semantics based on these considerations. These are just a few examples from an exciting emerging framework.

In terms of linguistic subdisciplines, morphology is one of the most important areas of linguistic research. How the basic units of meaning, morphemes, combine to create meaningful structures forms the cornerstone of language. In fact, the concept of syntax, that played such a significant role in generative linguistics, is less applicable in non-isolating or polysynthetic languages like Turkish or Yupik in which what we think of as sentences (strings of separate words) are not easily characterisable. In the tradition of the philosophy of linguistics, Dupre (2022) utilises morphological theory to challenge mainstream metasemantic accounts of reference, particularly Kripke-style arguments about reference chains based on individual word genealogies.

---

[34] https://link.springer.com/collections/cgicihfcie.

Bromberger (1989) was one of the first philosophers of linguistics to deploy the philosophy of science, specifically scientific modelling, to the ontology of linguistics. His framework for applying what he called the 'modelling condition' to linguistic types is based on artefactual models in science explicitly and what kinds of transferrable questions we can ask of them. From this platform, Bromberger and Halle (1989) launched their philosophy of phonology, highlighting the field's unique scientific features, hierarchy without infinity, and empirical grounding in terms of physical phonetic elements. Carr (1990), another philosopher of phonology, presents a case for an autonomous realist interpretation of linguistics based entirely off of Popper's falsification framework. His view opposes instrumentalism in a way similar to a recent article by Reiss (2024). Reiss mounts a defence of 'armchair linguistics' as a robust, empirical science despite its disconnection from laboratory experiments or statistical analysis, again with phonology (and syntax) as subject matter.

We have already mentioned metasemantics a few times in this Element. But it should be noted that it is a space in which the philosophy of science has been specifically harnessed in the exploration of foundational questions of meaning and its scientific study. Ball and Rabern (2018) provide a comprehensive volume with excellent chapters on various aspects of the study of meaning. The previously mentioned Yalcin (2018) (on model-based science in semantics) is included there. But Ball (2018) also offers a more fine-grained measurement-theoretic approach to semantic modelling and gradability.

Perhaps the most resources from the philosophy of science has been directed at the principle of compositionality in semantics. This central organising principle states that complex meaning is built up from (or a function of) atomic meaning and their syntax (Montague, 1970; Partee, 1984). But it is unclear whether or not this principle applies at the psychological, processing, or more metatheoretic level. And contraventions abound, leading in some cases to radical departures from standard linguistic theory (Croft, 2001; Goldberg, 2015). Inspired by a core methodological principle in the philosophy of science, Szabó (2012) argues that compositionality is an inference to the best explanation, a practice common in scientific explanation. He acknowledges the counter examples but maintains that compositionality is a methodological requirement for the construction of explanatory models of language.

The methodology of formal, model-theoretic semantics has even inspired spin-offs. The recent 'supersemantics' movement aims to extend the remit of semantics to iconic phenomena like emojis, dance, and pictures (Patel-Grosz et al., 2023). If you will recall from Section 3.1.3, much linguistic ire has been generated over the extension of formal principles to semantic data. Sometimes, this can feel a bit like applying quantum mechanics to macro-level phenomena.

Why should we assume linguistic meaning phenomena exhibit the same structure as other semiotic systems (including that of non-humans)? On the flip side, we might question whether formal semantics is a successful theory of linguistic meaning to begin with (Stanton, 2020; Stokhof, 2013). This is not to deny the quality of this recent work but rather to insist that we tread lightly and maintain an informed stance for future extensions.

Applied linguistics is perhaps a philosophical blind spot. This might be because theoretical linguistics itself often made little impact outside of purely theoretical contexts. Consider translation studies. The nature of translation was central to the more anthropological approach to linguistics. This makes sense since the job of linguists often involved translating little known languages. Mathematical or formal linguistics changed this picture significantly. A linguist armed with a decent knowledge of logic and their own introspective judgements of their mother tongue became sufficient for the task of understanding language *simpliciter*. While UG seems to promote cross-linguistic study, Bach (2013) questions whether the quest for universals truly achieved this end or was stifled by the assumption that all languages are basically the same. Balashov (2020, 2022) delves into neglected issues of translation, specifically in light of neural network deep learning models and their tokenisation procedures. He also applies the extended mind hypothesis (Clark & Chalmers, 1998) to automated translation among other things. Haspelmath (2024) performs some pioneering work in the philosophy of typology in arguing that universals can be found with broad methodologies as opposed to deep ones (the kinds suggested by UG and generative linguistics). His general views have questioned the methodological universalist agenda in theoretical linguistics, given the apparent incongruity of certain linguistic categories across different languages.

A related discipline that receives less than its due in theoretical linguistics is psycholinguistics. In the early approaches, such as Fodor, Bever, and Garrett (1974), the goal was to locate posits of theoretical or formal linguistics within the real-time processing of language. Various attempts, such as the famous click location experiments to find phrase boundaries, and the derivational theory of complexity based on the assumed processing of transformations proved elusive. Psycholinguistics largely followed its own path since. Theoretical linguists neatly categorised the results of this subfield in terms of the psychology of performance as opposed to competence. Indeed, many psycholinguistic effects are rooted in performance systems (replete with memory limitations and processing load constraints). For example, H. H. Clark and Tree (2002) find systematic differences in speech fillers like *uh* (indicating minor delays) and *um* (for more hefty hesitation). These fillers offer the interpreter subtle cues to the forthcoming message from the speaker. In theoretical linguistics, these kinds of patterns and phenomena are often elided over as dysfluencies in their scientific idealisations. Pereplyotchik (2017) challenges the orthodoxy here. He links psycholinguistic results to computational theoretical approaches to language, succouring E-language perspectives for acquisition along the way. This is done to bring real-time processing (parsing) considerations back into the competence model of language.

This brings us to neurolinguistics. In my view, neurolinguistics offers the most promising collaboration for the future science of language. Not only is there a theory problem, similar to psychology, in which large amounts of data has amassed, blind to and independent of linguistic theory. But advanced techniques in neural mapping can also contribute to answers to some of the most pressing theoretical questions. Is language for thought or communication? Friederici et al. (2017) and others isolate syntax in a subset of the BA 44 area (in the furthermost ventral anterior portion) in which Merge generates an infinite array of hierarchical structures. But for them, language should not be equated with communication, while their neural architecture aims for maximum correspondence with the theoretical picture in generative linguistics. As we have seen, Gibson et al. (2021) use information theory and efficiency considerations as evidence of the fundamental role communication plays in shaping language. What about the link between mathematical cognition and language underlying much of the infinity discussion in Section 4? Are mathematical cognition and linguistic cognition linked? Neurolinguistic evidence seems to suggest otherwise. Fedorenko and Varley (2016) analyse the fMRI data from healthy and impaired individuals to show that arithmetic abilities are unaffected by severe aphasia and left-hemisphere damage to the language system, concluding 'that brain regions that support linguistic (including grammatical) processing are not needed for exact arithmetic' (Fedorenko & Varley, 2016, 5).

What about next word prediction and the relevance of LLMs to linguistic theory? Is this kind of lexical prediction truly unrelated to brain activity? Henderson et al. (2016) show, via eye-tracking and fMRI data on reading tasks, that cortical regions in the inferior frontal gyrus and left anterior superior temporal lobe are responsible for syntactic prediction similar to that found in LLMs. Ryskin and Nieuwland (2023) suggest that our prediction abilities take preceding context into account (in a way similar to that of a transformer model). Thus, attention mechanisms and positional encoding in transformers and predictive coding in humans might share some overlapping features. Of course, artificial neurons are collections of vectors of numbers, while human neurons are based on complex biochemistry. But this doesn't mean the former cannot model the latter, as we know from the scientific modelling literature.

Neurolinguistics, like neuroscience, usually trades in event-related potentials (ERPs) or electrical brain activity fluctuations of specific areas of the brain, especially the cerebral cortex. Positive (P) peaks and negative (N) troughs in activity after the onset of stimulus can be measured quite precisely (in milliseconds) during electroencephalogram (EEG) or magnetoencephalography (MEG) studies. N100 and P200 activities have been found to be associated with phonological activity. N400 ERP is generally associated with word meaning in context. Specifically, the 'N400 effect' is the modulation of a negative peaking (or trough) around 400 ms when semantic priming or contextual shifts are experienced (Lau, Phillips, & Poeppel 2008). The initial discovery emerged when participants were confronted with unexpected words in reading tasks. Researchers expected a P300 effect (associated with surprisal) but found robust evidence for a semantic effect instead (see Kutas & Federmeier, 2011).

What about syntax? Well, careful experiments reveal that P600 activity is present when syntactic anomalies are perceived or pure grammatical markers (like infinitives) are scanned (Hagoort, Brown, & Groothusen, 1993). While this research does seem to confirm a certain division of labour between syntactic and semantic processing, the relatively late (latency) of P600 in processing compared to the semantic case is illustrative of additional possibilities. For instance, it suggests that syntax might not be the autonomous, prior system assumed in generative grammar but an incremental constraint on semantic processing.[35] FMRI studies offer more fine-grained analysis and data than EEG does. They do not measure electrical or electromagnetic activity but blood oxygen flow in the brain. Moro (2016)'s experiments on impossible languages tested participants on stimulus in order to map where the blood flow was augmented as an indication of the region in which the processing was taking place. As Baggio (2022, 48) notes:

> In reality, fMRI data are smooth spatial maps that allow us to infer (indirectly and probabilistically) which regions have been more or less active in the seconds following the stimuli.

Besides being 'indirect', such methods establish circumstantial evidence in terms of correlations. Language also seems to be distributed across brain regions in a way that can frustrate simple mapping exercises like EEG. Nevertheless, mapping pattern activation provides useful information and possibly even connections with NLP and LLM research. Cauchetuex and King (2022) use both fMRI and MEG in a study to measure the potential extent to which LLMs converge on brain-like processing on language tasks (e.g. reading tasks).

---

[35] This is similar to the formal framework that dynamic syntacticians have motivated, see Kempson, Meyer-Viol, and Gabbay (2001).

According to their results, and quite surprisingly, the correspondence with brain processing seems to depend on the LLM's performance on masked word tasks.

For a slightly different level of analysis, Murphy (2023) focuses on brain rhythms rather than activation patterns as a nexus of language development in the brain. This is partly to overcome the mapping correlation problems and their ramifications for scientific explanation (Poeppel, 2012).[36] Our brain resonates at certain frequency ranges (e.g. $\delta, \alpha, \gamma,$ and $\theta$) called 'neural oscillations', which 'reflect synchronized fluctuations in neuronal excitability' (Benítez-Burraco & Murphy, 2019, 1). In Murphy (2023), the approach is to reconstruct various theoretical linguistic operations and constructs (like Merge) within the 'oscillome'. Brain-imaging studies that focus exclusively on anatomical analysis are limited in terms of accessing functional complexity in the brain, some of which can be captured by appreciating the rhythms generated in different cortical and subcortical tissue. Even compositional structure can be approximated with 'phase synchronisation' which is 'consistent phase coupling between two neuronal signals oscillating at a given frequency' (Murphy, 2023, 67). His findings and modelling are fascinating and approach a realisation of a kind of 'neural syntax' based largely in the subcortex.

The proposal is not just that neurolinguistics can supplement existing methods or modes of evidence but that linguistic theory can corral the evidence into more unified explanatory frameworks. This would go some way to both justifying posits of theory and collating heaps of unanalysed data, potentially even buttressing scientific realism in the philosophy of linguistics by showing us where nature's real linguistic joints were hiding all along.

Moreover, in terms of the naturalistic philosophy of science advocated by Machery (2017), neurolinguistics provides a platform from which to turn abstract theoretical questions into concrete testable hypotheses. Undoubtedly, causation versus correlation issues beset the field and its techniques (even oscillatory ones). But another useful tool from the philosophy of scientific modelling is multiple models idealisation in which distinct models are used to capture convergence and interrelation (Weisberg, 2007). Yan and Hricko (2017) detail such an approach for cognitive neuroscience involving neural hubs and offer a structural realist take on how to map functional networks to anatomical ones in the brain.

---

[36] Again, connections to the philosophy of science are all over the place. This time the philosophy of applied mathematics/sciences spends a lot of time improving on the simple mapping account of mathematical application (Bueno & Colyvan, 2011b; Bueno, French, & Ladyman, 2002; Pincock, 2007; Wigner, 1960).

There is so much more to explore with the philosophies of science and linguistics as our guides. Language is perhaps one of the most complex phenomena in the natural world. Understanding it is sometimes call 'the hardest problem in science'. Such claims are difficult to assess but the philosophy of linguistics does hint at some reasons for the perplexity, given that it invites psychology, sociology, biology, neurobiology, information theory, computer science, mathematics, and physics to its careful study, with room to spare! This might make the philosophy of linguistics one of the most elaborate and sophisticated philosophies of science out there, implicating a wide range of other fields, both theoretical and empirical and teaching us something about the nature of the scientific enterprise in the process.

# References

Allott, N., Lohndal, T., & Rey, G. (2021a). Chomsky's 'Galilean' explanatory style. In N. Allott, T. Lohndal, & G. Rey (Eds.), *A Companion to Chomsky* (pp. 517–528). Oxford: Wiley-Blackwell.

Allott, N., Lohndal, T., & Rey, G. (Eds.). (2021b). *A Companion to Chomsky*. Hoboken, NJ: Wiley-Blackwell. https://doi.org/10.1002/9781119598732.

Allott, N., & Smith, N. (2021). Chomsky and Fodor on modularity. In N. Allott, T. Lohndal, & G. Rey (Eds.), *A Companion to Chomsky* (pp. 529–543). Hoboken, NJ: Wiley Blackwell.

Alter, S. G. (1999). *Darwinism and the linguistic image: Language, race, and natural theology in the nineteenth century*. Baltimore, MD: Johns Hopkins University Press.

Alter, S. G. (2013). Darwin and language. In M. Ruse (Ed.), *The Cambridge encyclopedia of Darwin and evolutionary thought* (pp. 182–187). Cambridge: Cambridge University Press.

Ananthaswamy, A. (2024). *Why machines learn: The elegant math behind modern AI*. New York: Dutton.

Anderson, S. R., & Lightfoot, D. W. (2002). *The language organ: Linguistics as cognitive physiology*. Cambridge: Cambridge University Press.

Ankeny, R., & Leonelli, S. (2011). What's so special about model organisms? *Studies in History and Philosophy of Science Part A*, *41*, 313–323.

Ankeny, R., & Leonelli, S. (2020). *Model organisms*. Cambridge: Cambridge University Press.

Appiah, K. A. (2017). *As if: Idealization and ideas*. Cambridge, MA: Harvard University Press.

Bach, E. (2013). Linguistic universals and particulars. In F. Kiefer, M. Ladányi, & P. Siptár (Eds.), *Eight decades of general linguistics* (pp. 489–502). Leiden: Brill.

Baggio, G. (2020). Epistemic transfer between linguistics and neuroscience: Problems and prospects. In R. M. Nefdt, C. Klippi, & B. Karstens (Eds.), *The philosophy and science of language* (pp. 279–300). Cham: Palgrave Macmillan.

Baggio, G. (2022). *Neurolinguistics*. Cambridge, MA: The MIT Press.

Baggio, G., & Murphy, E. (2024). *On the referential capacity of language models: An internalist rejoinder to Mandelkern & Linzen*. https://arxiv.org/abs/2406.00159.

Balari, S., Benítez-Burraco, A., Longa, V. M., & Lorenzo, G. (2013). The fossils of language: What are they? who has them? how did they evolve? In K. K. Grohmann & C. Boeckx (Eds.), *The Cambridge Handbook of Biolinguistics* (pp. 489–523). Cambridge: Cambridge University Press.

Balari, S., & González, G. L. (2013). *Computational phenotypes: Towards an evolutionary developmental biolinguistics*. Oxford: Oxford University Press.

Balari, S., & Lorenzo, G. (2009). Computational phenotypes: Where the theory of computation meets evo-devo. *Biolinguistics, 3*(1), 2–60.

Balashov, Y. (2020). The translator's extended mind. *Minds and Machines, 30*, 349–383.

Balashov, Y. (2022). The boundaries of meaning: A case study in neural machine translation. *Inquiry*, 1–34.

Ball, D. (2018). Semantics as measurement. In D. Ball & B. Rabern (Eds.), *The science of meaning: Essays on the metatheory of natural language semantics* (pp. 381–410). Oxford: Oxford University Press.

Ball, D., & Rabern, B. (Eds.). (2018). *The science of meaning: Essays on the metatheory of natural language semantics*. Oxford: Oxford University Press.

Baroni, M. (2022). On the proper role of linguistically oriented deep net analysis in linguistic theorising. In S. Lappin & J.-P. Bernardy (Eds.), *Algebraic structures in natural language* (pp. 1–16). Boca Raton, FL: CRC Press.

Becker, A. (2018). *What is real?: The unfinished quest for the meaning of quantum physics*. New York: Basic Books.

Behme, C. (2014). A 'Galilean' science of language. *Journal of Linguistics, 50*, 671–704.

Bender, E. M., & Koller, A. (2020, July). Climbing towards NLU: On meaning, form, and understanding in the age of data. In D. Jurafsky, J. Chai, N. Schluter, & J. Tetreault (Eds.), *Proceedings of the 58th annual meeting of the association for computational linguistics* (pp. 5185–5198). Online: Stroudsburg PA: Association for Computational Linguistics.

Benítez-Burraco, A., & Murphy, E. (2019). Why brain oscillations are improving our understanding of language. *Frontiers in Behavioral Neuroscience, 13*, 190.

Berwick, R. C., & Chomsky, N. (2016). *Why only us: Language and evolution*. Cambridge, MA: MIT Press.

Berwick, R. C., Friederici, A. D., Chomsky, N., & Bolhuis, J. J. (2013). Evolution, brain, and the nature of language. *Trends in Cognitive Sciences, 17*(2), 89–98.

Bever, T. G. (2013). The biolinguistics of language universals: The next years. In M. Sanz, I. Laka, & M. K. Tanenhaus (Eds.), *Language down the garden path: The cognitive and biological basis for linguistic structures* (pp. 385–405). Oxford: Oxford University Press.

Bever, T. G. (2021). How cognition came into being. *Cognition, 213*, 104761.

Bianchi, S. D. (2019). Combining finite and infinite elements: Why do we use infinite idealizations in engineering? *Synthese, 196*, 1733–1748.

Bickerton, D. (1975). *Dynamics of a creole system*. Cambridge: Cambridge University Press.

Bickerton, D. (2014). *More than nature needs: Language, mind, and evolution*. Cambridge, MA: Harvard University Press.

Bloomfield, L. (1936). Language or ideas? *Language, 12*, 89–95.

Boden, M. A. (2016). *AI: Its nature and future*. Oxford: Oxford University Press.

Boeckx, C. (2006). *Linguistic minimalism: Origins, concepts, methods, and aims*. Oxford: Oxford University Press.

Boeckx, C. (2010). Linguistic minimalism. In B. Heine & H. Narrog (Eds.), *The Oxford Handbook of Linguistic Analysis* (pp. 485–505). Oxford: Oxford University Press.

Boeckx, C., & Grohmann, K. K. (2007). The biolinguistics manifesto. *Biolinguistics, 1*(1–2), 1–8.

Boroditsky, L., Schmidt, L., & Phillips, W. (2003). Sex, syntax, and semantics. In D. Gentner & S. Goldin-Meadow (Eds.), *Language in mind: Advances in the study of language and cognition* (pp. 61–79). Cambridge, MA: The MIT Press.

Botha, R. P. (1982). On 'the Galilean style' of linguistic inquiry. *Lingua, 58*, 1–50.

Bromberger, S. (1989). Types and tokens in linguistics. In A. George (Ed.), *Reflections on chomsky* (pp. 58–90). Oxford: Basil Blackwell.

Bromberger, S., & Halle, M. (1989). Why phonology is different. *Linguistic Inquiry, 20*(1), 51–70.

Bueno, O., & Colyvan, M. (2011a). An inferential conception of the application of mathematics. *Noûs, 45*(2), 345–374.

Bueno, O., & Colyvan, M. (2011b). An inferential conception of the application of mathematics. *Noûs, 45*(2), 345–374.

Bueno, O., & French, S. (2018). *Applying mathematics: Immersion, inference, interpretation*. Oxford: Oxford University Press.

Bueno, O., French, S., & Ladyman, J. (2002). On representing the relationship between the mathematical and the empirical. *Philosophy of Science, 69*, 497–518.

Carr, P. (1990). *Linguistic realities: An autonomist metatheory for the generative enterprise.* Cambridge: Cambridge University Press.

Caucheteux, C., & King, J.-R. (2022). Brains and algorithms partially converge in natural language processing. *Communications Biology, 5,* 134.

Chaabouni, R., Kharitonov, E., Dupoux, E., & Baroni, M. (2019). Anti-efficient encoding in emergent communication. *ArXiv, abs/1905.12561.*

Chater, N., Clark, A., Goldsmith, J. A., & Perfors, A. (2015). *Empiricism and Language Learnability.* Oxford: Oxford University Press.

Chemla, E., & Nefdt, R. M. (2024). *No such thing as a general learner: Language models and their dual optimization.* https://arxiv.org/abs/2408.09544.

Chomsky, N. (1955/1975). *The logical structure of linguistic theory.* New York: Plenum Press.

Chomsky, N. (1956). Three models for the description of language. *IRE Transactions on Information Theory, 2*(3), 113–124.

Chomsky, N. (1957). *Syntactic structures.* The Hague: Mouton.

Chomsky, N. (1959a). On certain formal properties of grammars. *Information and Control, 1,* 91–112.

Chomsky, N. (1959b). Review of Skinner's *Verbal Behavior. Language, 35,* 26–58.

Chomsky, N. (1963). Formal properties of grammars. In R. D. Luce, R. R. Bush, & E. Galanter (Eds.), *Handbook of mathematical psychology II* (pp. 323–418). New York: Wiley.

Chomsky, N. (1965). *Aspects of the Theory of Syntax.* Cambridge, MA: MIT Press.

Chomsky, N. (1966). *Cartesian linguistics: A chapter in the history of rationalist thought.* New York: Harper & Row.

Chomsky, N. (1981). *Lectures on government and binding.* Dordrecht: Foris.

Chomsky, N. (1986). *Knowledge of language: Its nature, origin and use.* Westport, CT: Praeger.

Chomsky, N. (1991a). Linguistics and adjacent fields: A personal view. In A. Kasher (Ed.), *The Chomskyan Turn* (pp. 3–25). Oxford: Blackwell.

Chomsky, N. (1991b). Linguistics and cognitive science: Problems and mysteries. In A. Kasher (Ed.), *The Chomskyan Turn* (pp. 26–53). Oxford: Blackwell.

Chomsky, N. (1993). *A minimalist program for linguistic theory* (Tech. Rep. No. 1). MIT, Department of Linguistics and Philosophy.

Chomsky, N. (1995a). Language and nature. *Mind, 104,* 1–61.

Chomsky, N. (1995b). *The minimalist program.* Cambridge, MA: MIT Press.

Chomsky, N. (2000). *New horizons in the study of language and mind.* New York: Cambridge University Press.

Chomsky, N. (2002). *On nature and language.* Cambridge: Cambridge University Press.

Chomsky, N. (2005). Three factors in language design. *Linguistic Inquiry, 36*(1), 1–22.

Chomsky, N. (2008). On phases. In R. Freidin, C. P. Otero, M. L. Zubizarreta, S. J. Keyser (Eds.), *Foundational issues in linguistic theory: Essays in honor of Jean-Roger Vergnaud* (pp. 133–166). Cambridge MA: The MIT Press.

Chomsky, N. (2013). Lecture 1: What is language? *Journal of Philosophy, 110*(12), 645–662.

Chomsky, N. (2017). Two notions of modularity. In R. G. De Almeida & L. R. Gleitman (Eds.), *On concepts, modules, and language: Cognitive science at its core* (pp. 25–40). Oxford: Oxford University Press.

Chomsky, N. (2023). *Conversations with Tyler: Noam Chomsky.* https://conversationswithtyler.com/episodes/noam-chomsky/ (Podcast).

Chomsky, N., & McGilvray, J. (2012). *The science of language: Interviews with James McGilvray.* Cambridge: Cambridge University Press.

Chomsky, N., Roberts, I., & Watumull, J. (2023). Noam chomsky: The false promise of ChatGPT. *The New York Times.* www.nytimes.com/2023/03/08/opinion/noam-chomsky-chatgpt-ai.html.

Chomsky, N., Seely, T. D., Berwick, R. C. et al. (2023). *Merge and the strong minimalist thesis.* Cambridge: Cambridge University Press.

Cichy, R. M., & Kaiser, D. (2019). Deep neural networks as scientific models. *Trends in Cognitive Sciences, 23*(4), 305–317.

Clark, A., & Chalmers, D. (1998). The extended mind. *Analysis, 58*(1), 7–19.

Clark, A., & Lappin, S. (2012). Computational learning theory and language acquisition. In R. Kempson, T. Fernando, & N. Asher (Eds.), *Philosophy of linguistics* (pp. 445–475). Amsterdam: North-Holland.

Clark, H. H., & Tree, J. E. F. (2002). Using uh and um in spontaneous speaking. *Cognition, 84*(1), 73–111.

Collins, J. (2023a). Generative linguistics: 'Galilean style'. *Language Sciences, 100*, 101585. https://doi.org/10.1016/j.langsci.2023.101585.

Collins, J. (2023b). Internalist priorities in a philosophy of words. *Synthese, 201*(110), 1–33.

Collins, C., & Stabler, E. (2016). A formalization of minimalist syntax. *Syntax, 19*(1), 43–78.

Cowart, W. (1997). *Experimental syntax: Applying objective methods to sentence judgments.* Thousand Oaks, CA: Sage.

Croft, W. (2001). *Radical construction grammar: Syntactic theory in typological perspective*. Oxford: Oxford University Press.
Dabrowska, E. (2015). What exactly is universal grammar, and has anyone seen it? *Frontiers in Psychology, 6*, 1-17, Article 852.
Darwin, C. (1871). *The descent of man, and selection in relation to sex*. London: John Murray.
Dediu, D., & Levinson, S. C. (2013). On the antiquity of language: The reinterpretation of Neandertal linguistic capacities and its consequences. *Frontiers in Psychology, 4*, 397, 1-17.
Dekker, P. (2012). *Dynamic semantics*. Netherlands: Springer Verlag.
Dennett, D. C. (1981). *The intentional stance*. Cambridge, MA: MIT Press.
Dennett, D. C. (1991). Real patterns. *Journal of Philosophy, 88*(1), 27–51.
Dennett, D. C. (2017). *From bacteria to bach and back*. New York: W.W. Norton.
Deutscher, G. (2010). *Through the language glass: Why the world looks different in other languages*. New York: Metropolitan Books.
Devitt, M. (2005). Realism/anti-realism. In M. Curd & S. Psillos (Eds.), *The Routledge companion to philosophy of science* (pp. 224–235). London: Routledge.
Devitt, M. (2006). *Ignorance of language*. Oxford: Clarendon Press.
Douglas, H. (2009). Reintroducing prediction to explanation. *Philosophy of Science, 76*(4), 444–463.
Dupre, G. (2021). (What) can deep learning contribute to theoretical linguistics? *Minds and Machines, 31*, 617–635.
Dupre, G. (2022). Reference and morphology. *Philosophy and Phenomenological Research, 106*(3), 655–676.
Dupre, G. (2024). Acquiring a language vs. inducing a grammar. *Cognition, 247*(C), 1-12, 105771.
Easwaran, K., Hájek, A., Mancosu, P., & Oppy, G. (2023). Infinity. In E. N. Zalta & U. Nodelman (Eds.), *The Stanford Encyclopedia of Philosophy* (Winter 2023 ed.). Palo Alto: Metaphysics Research Lab, Stanford University.
Egré, P. (2015). Explanation in linguistics. *Philosophy Compass, 10*(7), 451–462.
Evans, N., & Levinson, S. C. (2009). The myth of language universals: Language diversity and its importance for cognitive science. *Behavioral and Brain Sciences, 32*(5), 429–448.
Everaert, M. B., Huybregts, M. A., Chomsky, N., Berwick, R. C., & Bolhuis, J. J. (2015). Structures, not strings: Linguistics as part of the cognitive sciences. *Trends in Cognitive Sciences, 19*(12), 729–743.

Everett, D. (2005). Cultural constraints on grammar and cognition in Pirahã: Another look at the design features of human language. *Current Anthropology*, *46*(4), 621–646. https://doi.org/10.1086/431525.

Everett, D. (2012). What does Pirahã grammar have to teach us about human language and the mind? *WIREs Cognitive Science*, *3*(6), 555–563.

Everett, D. (2017). *How language began: The story of humanity's greatest invention*. New York: W.W. Norton.

Featherston, S. (2007). Data in generative grammar: The stick and the carrot. *Theoretical Linguistics*, *33*, 269–318.

Fedorenko, E., Piantadosi, S. T., & Gibson, E. A. F. (2024). Language is primarily a tool for communication rather than thought. *Nature*, *630*, 575–586.

Fedorenko, E., & Varley, R. (2016, April). Language and thought are not the same thing: Evidence from neuroimaging and neurological patients. *Annals of the New York Academy of Sciences*, *1369*(1), 132–153.

Fletcher, S. C., Palacios, P., Ruetsche, L. et al. (2019). Infinite idealizations in science: An introduction. *Synthese*, *196*, 1657–1669.

Fodor, J. A. (1974). Special sciences (or: The disunity of science as a working hypothesis). *Synthese*, *28*(2), 97–115.

Fodor, J. A. (1983). *The modularity of mind: An essay on faculty psychology*. Cambridge, MA: MIT Press.

Fodor, J. A., Bever, T. G., & Garrett, M. F. (1974). *The psychology of language: An introduction to psycholinguistics and generative grammar*. New York: McGraw-Hill.

Frank, M. C. (2023). Bridging the data gap between children and large language models. *Trends in Cognitive Sciences*, *27*(11), 990–992.

Franke, M. (2013). Game theoretic pragmatics. *Philosophy Compass*, *8*(3), 269–284.

Frankish, K., & Ramsey, W. M. (Eds.). (2012). *The Cambridge companion to cognitive science*. Cambridge: Cambridge University Press.

Freidin, R. (2013). Noam Chomsky's contribution to linguistics: A sketch. In K. Allan (Ed.), *The Oxford handbook of the history of linguistics* (pp. 438–467). Oxford: Oxford University Press.

French, S. (2014). *The philosophy of science: Key concepts*. London: Bloomsbury Academic.

French, S. (2016). *Philosophy of science: Key concepts*. London: Bloomsbury Academic.

Friederici, A. D., Chomsky, N., Berwick, R. C., Moro, A., & Bolhuis, J. J. (2017). Language, mind and brain. *Nature Human Behaviour*, *1*(10), 713–722. https://doi.org/10.1038/s41562-017-0184-4.

Gardner, H. (1985a). *The mind's new science: A history of the cognitive revolution*. New York: Basic Books.
Gardner, H. (1985b). *The mind's new science: A history of the cognitive revolution*. New York: Basic Books.
Gibson, E., Futrell, R., Piantadosi, S. T. et al. (2019). How efficiency shapes human language. *Trends in Cognitive Sciences*, *23*(5), 389–407.
Gibson, E., Futrell, R., Piantadosi, S. T. et al. (2021). How efficiency shapes human language. *Trends in Cognitive Sciences*, *25*(5), 389–407.
Gibson, E., & Poliak, M. (Eds.). (2024). *From fieldwork to linguistic theory* (No. 15). Berlin: Language Science Press.
Giere, R. N. (1988). *Explaining science: A cognitive approach*. Chicago, IL: University of Chicago Press.
Glanzberg, M. (2014). Explanation and partiality in semantic theory. In A. Burgess & B. Sherman (Eds.), *Metasemantics: New essays on the foundations of meaning* (pp. 135–167). Oxford: Oxford University Press.
Godfrey-Smith, P. (2003). *Theory and reality: An introduction to the philosophy of science*. Chicago, IL: University of Chicago Press.
Godfrey-Smith, P. (2006). The strategy of model-based science. *Biology and Philosophy*, *21*(5), 725–740.
Godfrey-Smith, P. (2016). *Other minds: The octopus, the sea, and the deep origins of consciousness*. New York: Farrar, Straus and Giroux.
Gold, E. M. (1967). Language identification in the limit. *Information and Control*, *10*(5), 447–474.
Goldberg, A. E. (2015). Compositionality. In E. Dabrowska & D. Divjak (Eds.), *The Oxford handbook of construction grammar* (pp. 473–492). Oxford: Oxford University Press.
Goldsmith, J. A. (2001). Unsupervised learning of the morphology of a natural language. *Computational Linguistics*, *27*(2), 153–198.
Goldsmith, J. A., & Laks, B. (2019). *Battle in the mind fields*. Chicago, IL: University of Chicago Press.
Gopnik, M. (1990). Feature-blind grammar and dysphasia. *Nature*, *344*, 715.
Graf, T. (2022). Subregular linguistics: Bridging theoretical linguistics and formal grammar. *Theoretical Linguistics*, *48*(3–4), 145–184. https://doi.org/10.1515/tl-2022-2037.
Grice, H. P. (1975). Logic and conversation. In P. Cole & J. Morgan (Eds.), *Syntax and semantics* (Vol. 3, pp. 41–58). New York: Academic Press.
Hagoort, P., Brown, C., & Groothusen, J. (1993). The syntactic positive shift (SPS) as an ERP measure of syntactic processing. *Language and Cognitive Processes*, *8*(4), 439–483.

Harris, R. A. (2021). *The linguistics wars: Chomsky, Lakoff, and the battle over deep structure* (2nd ed.). New York: Oxford University Press.

Haspelmath, M. (2024). Breadth versus depth: Theoretical reasons for system-independent comparison of languages. In R. M. Nefdt, G. Dupre, & K. Stanton (Eds.), *Oxford Handbook of the Philosophy of Linguistics*. Oxford: Oxford University Press.

Hasson, U., Nastase, S. A., & Goldstein, A. (2020). Direct fit to nature: An evolutionary perspective on biological and artificial neural networks. *Neuron, 105*(3), 416–434.

Hauser, M., Chomsky, N., & Fitch, T. (2002). The faculty of language: What is it, who has it, and how did it evolve? *Science, 298*(22), 1569–1579.

Heim, I. (1982). *The semantics of definite and indefinite noun phrases* (Unpublished doctoral dissertation). University of Massachusetts, Amherst. (File Change Semantics)

Hellman, G. (1989). *Mathematics without numbers: Towards a modal-structural interpretation*. Oxford: Clarendon Press.

Hempel, C. G., & Oppenheim, P. (1948). Studies in the logic of explanation. *Philosophy of Science, 15*, 135–175.

Henderson, J. M., Choi, W., Lowder, M. W., & Ferreira, F. (2016). Language structure in the brain: A fixation-related fMRI study of syntactic surprisal in reading. *NeuroImage, 132*, 293–300.

Hesse, M. B. (1963). *Models and analogies in science*. Notre Dame, IN: University of Notre Dame Press.

Hintikka, J. (1999). The emperor's new intuitions. *Journal of Philosophy, 96*(3), 127–147.

Hinzen, W., & Sheehan, M. (2015). *The philosophy of universal grammar*. Oxford: Oxford University Press.

Hinzen, W., & Uriagereka, J. (2006). On the metaphysics of linguistics. *Erkenntnis, 65*(1), 71–96.

Horn, L., & Kecskes, I. (2013). Pragmatics, discourse, and cognition. In S. Anderson, J. Moeschler, & A. Reboul (Eds.), *The language-cognition interface* (pp. 353–375). Geneva: Librairie Droz.

Hsu, A. S., & Chater, N. (2010). The logical problem of language acquisition: A probabilistic perspective. *Cognitive Science, 34*(6), 972–1016.

Hsu, A. S., Chater, N., & Vitányi, P. M. (2011). The probabilistic analysis of language acquisition: Theoretical, computational, and experimental analysis. *Cognition, 120*(3), 380–390.

Huck, G. J., & Goldsmith, J. A. (1995). *Ideology and linguistic theory: Noam chomsky and the deep structure debates*. London: Routledge.

Hughes, R. I. G. (1997). Models and representation. *Philosophy of Science, 64*, S325–S336.
Huybregts, M. A. C. (2019). Infinite generation of language unreachable from a stepwise approach. *Frontiers in Psychology, 10*, 1-9.
Itkonen, E. (2001). Concerning the philosophy of phonology. *Puhe ja kieli, 21*, 3–11.
Jackendoff, R. (2002). *Foundations of language: Brain, meaning, grammar, evolution.* Oxford: Oxford University Press.
Jackendoff, R. (2018). Representations and rules in language. In B. Huebner (Ed.), *The Philosophy of Daniel Dennett* (pp. 95–126). Oxford: Oxford University Press.
Jackendoff, R., & Wittenberg, E. (2014). What you can say without syntax: A hierarchy of grammatical complexity. In F. Newmeyer & L. Preston (Eds.), *Measuring linguistic complexity* (pp. 65–82). Oxford: Oxford University Press.
Jackson, B. B. (2020). Model-theoretic semantics as model-based science. *Synthese, 199*(1–2), 3061–3081.
Jacobson, P. (1999). Sapir's legacy and the science of linguistics. *Journal of Linguistics, 35*(2), 225–240.
Jacobson, P. (2012). Direct compositionality. In W. Hinzen, E. Machery, & M. Werning (Eds.), *The oxford handbook of compositionality* (pp.109–128). Oxford: Oxford University Press. https://doi.org/10.1093/oxfordhb/9780199541072.013.0005 (Online edition, accessed 16 March 2025)
Jäger, G., & Rogers, J. (2012). Formal language theory: Refining the Chomsky Hierarchy. *Philosophical Transactions of the Royal Society B: Biological Sciences, 367*(1598), 1956–1970.
Jech, T. (2005). Set theory. *Bulletin of Symbolic Logic, 11*(2), 243–245.
Jelinek, F. (1988). Applying information theoretic methods: Evaluation of grammar quality. In *Proceedings of the workshop on evaluation of natural language processing systems.* Wayne, PA.
Jerne, N. K. (1985). The generative grammar of the immune system. *Bioscience Reports, 5*(6), 439–451.
Johnson, K. (2007). The legacy of methodological dualism. *Mind and Language, 22*(4), 366–401. https://doi.org/10.1111/j.1468-0017.2007.00313.x
Johnson, K. (2015). Notational variants and invariance in linguistics. *Mind & Language, 30*(2), 162–186.
Johnson, M., & Lakoff, G. (2002). Why cognitive linguistics requires embodied realism. *Cognitive Linguistics, 13*(3), 245–263.

Joseph, J. E. (1999). How structuralist was 'American Structuralism'? *Henry Sweet Society for the History of Linguistic Ideas Bulletin, 33*(1), 23–28.

Joshi, A. K. (1985). How much context sensitivity is required to provide reasonable structural descriptions: Tree adjoining grammars. In D. Dowty, L. Karttunen, & A. Zwicky (Eds.), *Natural language processing: Psycholinguistic, computational and theoretical perspectives* (pp. 206–250). Cambridge: Cambridge University Press.

Kallini, J., Papadimitriou, I., Futrell, R., Mahowald, K., & Potts, C. (2024). *Mission: Impossible language models.* https://arxiv.org/abs/2401.06416.

Kaplan, D. (1990). Words. *Aristotelian Society Supplementary Volume, 64*(1), 93–119.

Katz, J. (1972). *Linguistic philosophy: The underlying reality of language and its philosophical import.* East Melbourne: Allen and Unwin.

Katz, J. (1981). *Language and other abstract objects.* Maryland: Rowman & Littlefield.

Katzir, R. (2023, 12). Why large language models are poor theories of human linguistic cognition: A reply to Piantadosi. *Biolinguistics, 17*, 1-12, e13153.

Keiser, J. (2022). *Non-ideal foundations of language.* New York: Routledge.

Kempson, R. M., Meyer-Viol, W., & Gabbay, D. (2001). *Dynamic syntax: The flow of language understanding.* Oxford: Blackwell.

Kenneally, C. (2007). *The first word: The search for the origins of language.* New York: Viking.

Kincaid, H. (2008). Structural realism and the social sciences. *Philosophy of Science, 75*(5), 720–731.

Kitcher, P. (1989). Explanatory unification and the causal structure of the world. In P. Kitcher & W. Salmon (Eds.), *Scientific explanation* (pp. 410–505). Minneapolis, MN: University of Minnesota Press.

Knuuttila, T., & Merz, M. (2009). Understanding by modeling: An objectual approach. In H. W. de Regt, S. Leonelli, & K. Eigner (Eds.), *Scientific understanding: Philosophical perspectives* (pp. 146–168). Pittsburgh, PA: University of Pittsburgh Press.

Kodner, J., Payne, S., & Heinz, J. (2023). *Why linguistics will thrive in the 21st century: A reply to Piantadosi (2023).* https://arxiv.org/abs/2308.03228.

Korta, K., & Perry, J. (2008). Pragmatics. In K. Korta & J. Perry (Eds.), *Stanford Encyclopedia of Philosophy.* Palo Alto: CSLI.

Kragh, H. (2014). *The true (?) story of Hilbert's infinite hotel.* https://doi.org/arXiv:1403.0059.

Kretzschmar, W. A. J. (2015). *Language as a complex system.* Cambridge: Cambridge University Press.

Kuhn, T. S. (1962). *The structure of scientific revolutions* (1st ed.). Chicago, IL: University of Chicago Press. (Revised editions published in 1970 and 1996).

Kutas, M., & Federmeier, K. D. (2011). Thirty years and counting: Finding meaning in the N400 component of the event-related brain potential (ERP). *Annual Review of Psychology, 62*, 621–647.

Jäger, G., & Rogers, J. (2012) Formal language theory: Refining the Chomsky hierarchy. *Philosophical Transactions of the Royal Society B, 367*, 715–762.

Labov, W. (1969). Contraction, deletion, and inherent variability of the English copula. *Language, 45*, 715–762.

Ladyman, J. (1998). What is structural realism? *Studies in History and Philosophy of Science Part A, 29*(3), 409–424.

Ladyman, J., & Ross, D. (2007). *Everything must go: Naturalized metaphysics*. Oxford: Oxford University Press.

Lakoff, G. (1991). Cognitive versus generative linguistics: How commitments influence results. *Language & Communication, 11*(1/2), 53–62.

Lakoff, G., & Johnson, M. (1980). *Metaphors we live by*. Chicago, IL: University of Chicago Press.

Lakoff, G., & Johnson, M. (1999). *Philosophy in the flesh: The embodied mind and its challenge to western thought*. New York: Basic Books.

Landman, F. (1991). *Structures for semantics*. Netherlands: Springer Verlag.

Langendoen, D. T. (2010). Just how big are natural languages? In H. van der Hulst (Ed.), *Recursion and human language* (pp. 139–146). Berlin: De Gruyter Mouton.

Lappin, S. (2024). Assessing the strengths and weaknesses of large language models. *Journal of Logic, Language and Information, 33*, 9–20.

Lappin, S., Levine, R. D., & Johnson, D. E. (2000). The structure of unscientific revolutions. *Natural Language and Linguistic Theory, 18*(3), 665–671.

Lau, E., Phillips, C., & Poeppel, D. (2008). A cortical network for semantics: (de)constructing the N400. *Nature Reviews Neuroscience, 9*(12), 920–933.

Lenneberg, E. H. (1967). *Biological foundations of language*. New York: Wiley.

Levelt, W. J. (2008). *An introduction to the theory of formal languages and automata*. John Benjamins.

Lewis, D. K. (1969). *Convention: A philosophical study*. Amsterdam,: Cambridge, MA: Harvard University Press.

Lewis, D. K. (1975). Languages and language. In K. Gunderson (Ed.), *Minnesota studies in the philosophy of science* (pp. 3–35). Minnesota: University of Minnesota Press.

Lewis, D. K. (1979). Scorekeeping in a language game. *Journal of Philosophical Logic, 8*, 339–359.

Lewis, D. K. (1986). *On the plurality of worlds*. Oxford: Blackwell.

Lian, Y., Bisazza, A., & Verhoef, T. (2021). The effect of efficient messaging and input variability on neural-agent iterated language learning. *ArXiv, abs/2104.07637*.

Linnebo, Ø., & Shapiro, S. (2017). Actual and potential infinity. *Noûs, 53*(1), 160–191.

Linzen, T., & Baroni, M. (2021). Syntactic structure from deep learning [Journal Article]. *Annual Review of Linguistics, 7*, 195–212.

Lobina, D. J. (2017). *Recursion: A computational investigation into the representation and processing of language*. Oxford University Press.

Loewer, B. (2009). Why is there anything except physics? *Synthese, 170*(2), 217–233.

Ludlow, P. (2011). *Philosophy of generative grammar*. Oxford: Oxford University Press.

Machery, E. (2017). *Philosophy within its proper bounds*. Oxford: Oxford University Press.

Mallory, F. (2020). Linguistic types are capacity-individuated action-types. *Inquiry, 63*(9-10), 1123–1148.

Mallory, F. (2023). Why is generative grammar recursive? *Erkenntnis, 88*, 3097–3111.

Mallory, F. (2024). Generative linguistics and the computational level. *Croatian Journal of Philosophy, 24*(71), 195–218.

Mandelkern, M., & Linzen, T. (2024). *Do language models' words refer?* https://arxiv.org/abs/2308.05576.

Marantz, A. (2005). Generative linguistics within the cognitive neuroscience of language. *The Linguistic Review, 22*(2–4), 429–445.

Marcus, G. F. (1993). Negative evidence in language acquisition. *Cognition, 46*(1), 53–85.

Markus, I., & Bringmann, L. F. (2021). The theory crisis in psychology: How to move forward. *Perspectives on Psychological Science, 16*(4), 779–788.

Marr, D. (1982). *Vision: A computational investigation into the human representation and processing of visual information*. San Francisco: W.H. Freeman.

Martins, P. T., & Boeckx, C. (2019). Language evolution and complexity considerations: The no half-merge fallacy. *PLOS Biology*, *17*(11), 1–5, e3000389.

Matthews, P. (2001). *A short history of structural linguistics*. Cambridge: Cambridge University Press.

McWhorter, J. H. (2014). *The language hoax: Why the world looks the same in any language*. Oxford: Oxford University Press.

Mesthrie, R., & Nefdt, R. M. (in press). Sociolinguistics as deidealisation. In R. Nefdt, G. Dupre, & K. Stanton (Eds.), *The Oxford handbook of the philosophy of linguistics*. Oxford: Oxford University Press. (forthcoming)

Miller, P. (1999). *Strong generative capacity: The semantics of linguistic formalism*. Stanford: CSLI.

Miller, G. (2003). The cognitive revolution: A historical perspective. *Trends in Cognitive Sciences*, *7*(3), 141–144.

Miller, J. (2021a). A bundle theory of words. *Synthese*, *198*, 5731–5748.

Miller, J. (2021b). Words, species, and kinds. *Metaphysics*, *4*(1), 18–31.

Miller, J. T. M., & Hughes, T. J. (2023). *The Philosophy of Words, Synthese*, https://link.springer.com/collections/cgicihfcie.

Millière, R. (2025). Language models as models of language. In R. M. Nefdt, G. Dupre, & K. Stanton (Eds.), *Oxford Handbook of the Philosophy of Linguistics*. Oxford: Oxford University Press.

Millière, R., & Buckner, C. (2024). *A philosophical introduction to language models–part I: Continuity with classic debates*. https://arxiv.org/abs/2401.03910.

Millikan, R. G. (1984). *Language, thought, and other biological categories: New foundations for realism*. Cambridge, MA: MIT Press.

Millikan, R. G. (2003). In defense of public language. In L. M. Antony & N. Hornstein (Eds.), *Chomsky and his critics* (pp. 215–237). Oxford: Wiley-Blackwell.

Millikan, R. G. (2005). *Language: A biological model*. Oxford: Oxford: Clarendon Press.

Mitchell, J., & Bowers, J. S. (2020). Priorless recurrent networks learn curiously. In D. Scott, N. Bel, & C. Zong (Eds.), *Proceedings of the 28th international conference on computational linguistics* (pp. 5147–5158). Barcelona (Online): International Committee on Computational Linguistics.

Montague, R. (1970). Universal grammar. *Theoria*, *36*(3), 373–398.

Morgan, M. S. (2013). Nature's experiments and natural experiments in the social sciences. *Philosophy of the Social Sciences*, *43*(3), 341–357.

Morgan, M. S., & Morrison, M. (1999). *Models as mediators*. Cambridge: Cambridge University Press.

Moro, A. (2016). *Impossible languages*. Cambridge, MA: The MIT Press.

Moro, A., Greco, M., & Cappa, S. F. (2023). Large languages, impossible languages and human brains. *Cortex, 167*, 82–85.

Morrison, M. (2015). *Reconstructing reality: Models, mathematics, and simulations*. Oxford: Oxford University Press.

Mukerji, N. (2022). *The Human Mind through the Lens of Language*. New York: Bloomsbury Press.

Müller, S. (2018). *Grammatical theory: From transformational grammar to constraint-based approaches*. Berlin: Language Science Press.

Murphy, E. (2012). *Biolinguistics and philosophy: Insights and obstacles*. Lulu.com: Morrisville, NC: Lulu.

Murphy, E. (2023). *The oscillatory nature of language*. Cambridge: Cambridge University Press.

Nefdt, R. M. (2016). Scientific modelling in generative grammar and the dynamic turn in syntax. *Linguistics and Philosophy, 39*(5), 357–394.

Nefdt, R. M. (2018). Languages and other abstract structures. In C. Behme & M. Neef (Eds.), *Essays on linguistic realism* (pp. 139–184). Amsterdam: John Benjamins.

Nefdt, R. M. (2019a). Infinity and the foundations of linguistics. *Synthese, 196*(5), 1671–171.

Nefdt, R. M. (2019b). The ontology of words: A structural approach. *Inquiry: An Interdisciplinary Journal of Philosophy, 62*(8), 877–911.

Nefdt, R. M. (2020a). Formal semantics and applied mathematics: An inferential account. *Journal of Logic, Language & Information, 29*(2), 221–253.

Nefdt, R. M. (2020b). The role of language in the cognitive sciences. In R. Nefdt, C. Klippi, & B. Karstens (Eds.), *The philosophy and science of language* (pp. 235–256). Cham: Palgrave Macmillan.

Nefdt, R. M. (2021). Structural realism and generative linguistics. *Synthese, 199*(1–2), 3711–3737.

Nefdt, R. M. (2023a). *Language, science, and structure: A journey into the philosophy of linguistics*. New York: Oxford University Press.

Nefdt, R. M. (2023b). Motivating a scientific modeling continuum: The case of 'natural models' in the covid-19 pandemic. *Philosophy of Science, 90*(4), 880–900.

Nefdt, R. M. (2023b). Biolinguistics and biological systems: A complex systems analysis of language. *Biology and Philosophy, 38*(2), 1–42.

Nefdt, R. M. (2024). *The philosophy of theoretical linguistics: A contemporary outlook*. Cambridge: Cambridge University Press.

Nefdt, R. M., & Baggio, G. (2024a). Notational variants and cognition: The case of dependency grammar. *Erkenntnis*, (89), 2867–2897.

Nefdt, R. M., & Kac, M. B. (2025). Some thoughts on formalization in linguistics. In R. M. Nefdt, G. Dupre, & K. Stanton (Eds.), *The Oxford handbook of the philosophy of linguistics*. Oxford: Oxford University Press.

Nevins, A., Pesetsky, D., & Rodrigues, C. (2009). Pirahã exceptionality: A reassessment. *Language*, *85*, 355–404.

Newmeyer, F. J. (1986a). Has there been a 'Chomskyan revolution' in linguistics? *Language*, *62*(1), 1–18.

Newmeyer, F. J. (1986b). *Linguistic theory in America* (2nd ed.). Orlando: Academic Press.

Newmeyer, F. J. (1996). *Generative linguistics: An historical perspective* (1st ed.). London: Routledge.

Oppy, G. (2006). *Philosophical perspectives on infinity*. New York: Cambridge University Press.

Papineau, D. (2016). Teleosemantics. In D. L. Smith (Ed.), *How biology shapes philosophy: New foundations for naturalism* (pp. 95–120). Cambridge: Cambridge University Press.

Partee, B. H. (1984). Compositionality. In F. Landman & F. Veltman (Eds.), *Varieties of formal semantics* (pp. 281–311). Dordrecht: Foris.

Partee, B. H., ter Meulen, A., & Wall, R. E. (1990). *Mathematical methods in linguistics* (Vol. 30). Dordrecht: Springer.

Pasquiou, A., Lakretz, Y., Hale, J., Thirion, B., & Pallier, C. (2022). *Neural language models are not born equal to fit brain data, but training helps.* https://arxiv.org/abs/2207.03380.

Patel-Grosz, P., Mascarenhas, S., Chemla, E. et al. (2023). Super linguistics: An introduction. *Linguistics and Philosophy*, *46*, 627–692.

Pearl, L. (2021). Poverty of the stimulus without tears. *Language Learning and Development*, *18*(4), 415–454.

Pelletier, J., & Nefdt, R. M. (2025). *Linguistic relativity: A guide to past debates and future prospects*. New York: Oxford University Press.

Percival, W. K. (1976). The applicability of Kuhn's paradigms to the history of linguistics. *Language*, *52*(2), 285–294.

Pereplyotchik, D. (2017). *Psychosyntax: The nature of grammar and its place in the mind*. Cham: Springer Verlag.

Phillips, C., & Lasnik, H. (2003). Linguistics and empirical evidence: Reply to Edelman and Christiansen. *Trends in Cognitive Sciences*, *7*, 61–62.

Piantadosi, S. T. (2024). Modern language models refute Chomsky's approach to language. In E. Gibson & M. Poliak (Eds.), *From fieldwork to linguistic theory: A tribute to Dan Everett* (Vol. 15, pp. 353–414). Berlin: Language Science Press.

Pincock, C. (2007). A role for mathematics in the physical sciences. *Noûs, 41*, 253–275.

Planer, R. J., & Sterelny, K. (2021). *From signal to symbol: The evolution of language*. Cambridge, MA: The MIT Press.

Poeppel, D. (2012). The maps problem and the mapping problem: Two challenges for a cognitive neuroscience of speech and language. *Cognitive Neuropsychology, 29*(1–2), 34–55.

Poeppel, D., & Embick, D. (2005). Defining the relation between linguistics and neuroscience. In A. Cutler (Ed.), *Twenty-first century psycholinguistics: four cornerstones* (pp. 103–118). Mahwah, NJ: Lawrence Erlbaum Associates.

Popper, K. R. (1959). *The logic of scientific discovery*. London: Hutchinson.

Postal, P. M. (2009). The incoherence of Chomsky's 'biolinguistic' ontology. *Biolinguistics, 3*(1), 104–123.

Potts, C. (2024). Characterizing English preposing in pp constructions. *Journal of Linguistics*, 1–39.

Progovac, L. (2015). *Evolutionary syntax*. Oxford: Oxford University Press.

Progovac, L. (2016). A gradualist scenario for language evolution: Precise linguistic reconstruction of early human (and Neandertal) grammars. *Frontiers in Psychology, 7*, 1714.

Pullum, G. K. (2011). On the mathematical foundations of syntactic structures. *Journal of Logic, Language and Information, 20*(3), 277–296.

Pullum, G. K. (2013). The central question in comparative syntactic metatheory. *Mind and Language, 28*(4), 492–521.

Pullum, G. K., & Scholz, B. (2002). Empirical assessment of stimulus poverty arguments. *The Linguistic Review, 19*, 9–50.

Pullum, G. K., & Scholz, B. C. (2010). Recursion and the infinitude claim. In H. van der Hulst (Ed.), *Recursion and Human language* (pp. 111–138). Netherlands: De Gruyter Mouton.

Putnam, H. (1975). Explanation and reference. In H. Putnam (Ed.), *Mathematics, matter and method: Philosophical papers, volume 1* (pp. 196–214). Cambridge: Cambridge University Press.

Putnam, H. (1981). *Reason, truth, and history*. Cambridge: Cambridge University Press.

Pylyshyn, Z. W. (1984). *Computation and cognition*. Cambridge, MA: MIT press.

Quine, W. V. O. (1969). Epistemology naturalized. In *Ontological relativity and other essays* (pp. 69–90). New York: Columbia University Press.

Quine, W. V. O. (1972). Methodological reflections on current linguistic theory. In D. Davidson & G. Harman (Eds.), *Semantics of natural language* (pp. 442–454). Dordrecht: Reidel.

Rapaport, W. J. (2012). *Philosophy of computer science: An introduction to the issues and the literature* (Tech. Rep.). University at Buffalo, The State University of New York.

Rayo, A. (2019). *On the brink of paradox*. Cambridge, MA: MIT Press.

Reiss, C. (2024). Research methods in armchair linguistics. In R. Nefdt, G. Dupre, & K. Stanton (Eds.), *Oxford handbook of philosophy of linguistics*. Oxford: Oxford University Press. (Forthcoming)

Resnik, M. D. (1982). Mathematics as a science of patterns: Epistemology. *Noûs*, *16*(1), 95–105.

Rey, G. (2020). *Representation of language: Philosophical issues in a Chomskyan linguistics*. Oxford: Oxford University Press.

Richard, M. (2019). *Meanings as species*. Oxford: Oxford University Press.

Ross, J. R. (1967). *Constraints on variables in syntax* (Ph.D. Dissertation). Massachusetts Institute of Technology.

Ryskin, R., & Nieuwland, M. S. (2023). Prediction during language comprehension: What is next? *Trends in Cognitive Sciences*, *27*(11), 1032–1052.

Sampson, G. (2005). *The 'language instinct' debate*. London: Continuum Press.

Sapir, E. (1929). The status of linguistics as a science. *Language*, *5*(4), 207–214.

Savitch, W. J. (1993). Why it might pay to assume that languages are infinite. *Annals of Mathematics and Artificial Intelligence*, *8*(1–2), 17–25.

Schindler, S., Drożdżowicz, A., & Brøcker, K. (Eds.). (2020). *Linguistic intuitions: Evidence and method*. Oxford: Oxford University Press.

Schlenker, P. (2018). Visible meaning: Sign language and the foundations of semantics. *Theoretical Linguistics*, *44*(3–4), 123–208.

Scholz, B. C., Pelletier, F. J., Pullum, G. K., & Nefdt, R. M. (2022). Philosophy of linguistics. In E. N. Zalta (Ed.), *The Stanford encyclopedia of philosophy* (Fall ed.). Palo Alto: Metaphysics Research Lab, Stanford University. https://plato.stanford.edu/entries/linguistics/.

Schütze, C. T. (1996). *The empirical base of linguistics: Grammaticality judgments and linguistic methodology*. Chicago, IL: University of Chicago Press.

Searle, J. R. (1980). Minds, brains, and programs. *The Behavioral and Brain Sciences*, *3*(3), 417–457.

Sellars, W. (1953). Inference and meaning. *Mind*, *62*(247), 313–338.

Seuren, P. (2013). *From Whorf to Montague*. Oxford: Oxford University Press.

Shapiro, S. (1997). *Philosophy of mathematics: Structure and ontology*. New York: Oxford University Press. (Sometimes cited as *Mathematics as a Science of Patterns*).

Shea, N. (2018). *Representation*. Oxford: Oxford University Press.

Sinha, C. (2010). Cognitive linguistics, psychology, and cognitive science. In D. Geeraerts & H. Cuyckens (Eds.), *The Oxford Handbook of Cognitive Linguistics* (pp. 1–30). Oxford: Oxford University Press.

Skinner, B. (1957). *Verbal behavior*. Acton, MA: Copley.

Smith, N. (1994). Chomsky's revolution. *Nature*, *367*(6460), 521–522. https://doi.org/10.1038/367521a0.

Sperber, D., & Wilson, D. (1995). *Relevance: Communication and cognition* (2nd ed.). Oxford: Blackwell.

Sprouse, J., & Almeida, D. (2012). Assessing the reliability of textbook data in syntax: Adger's core syntax1. *Journal of Linguistics*, *48*(3), 609–652.

Stalnaker, R. (1976). Possible worlds. *Noûs*, *10*(1), 65–75. https://doi.org/10.2307/2214477.

Stalnaker, R. (1978). Assertion. *Syntax and Semantics (New York Academic Press)*, *9*, 315–332.

Stalnaker, R. (2002). Common ground. *Linguistics and Philosophy*, *25*, 701–721.

Stalnaker, R. (2014). *Context: Context and content*. Oxford: Oxford University Press.

Stanley, J., & Szabó, Z. G. (2000). On quantifier domain restriction. *Mind & Language*, *15*(2–3), 219–261.

Stanton, K. H. (2020). Linguistics and philosophy: Break up song. In R. M. Nefdt, C. Klippi, & B. Karstens (Eds.), *The philosophy and science of language* (pp. 315–333). Cham: Palgrave Macmillan.

Stanton, K. H. (2023). Composing words and non-words. *Synthese*, *202*(179), 1-28.

Steedman, M. (2017). The emergence of language. *Mind & Language*, *32*, 579–590.

Stokhof, M. (2013). Formal semantics and Wittgenstein. *The Monist*, *96*(2), 205–231.

Strevens, M. (2020). *The knowledge machine: How irrationality created modern science*. New York: Liveright.

Suppes, P. (1968). The desirability of formalization in science. *Journal of Philosophy*, *65*(20), 651–664.

Szabó, Z. G. (2012). The case for compositionality. In M. Werning, W. Hinzen, & E. Machery (Eds.), *The Oxford Handbook of Compositionality* (pp. 64–80). Oxford: Oxford University Press.

Taran, S., Adhikari, N. K. J., & Fan, E. (2021). Falsifiability in medicine: What clinicians can learn from Karl Popper. *Intensive Care Medicine, 47*(9), 1054–1056.

Tiede, H.-J., & Stout, L. N. (2010). Recursion, infinity, and modeling. In H. van der Hulst (Ed.), *Recursion and human language* (pp. 147–158). Berlin: De Gruyter Mouton.

Tomalin, M. (2006). *Linguistics and the formal sciences: The origins of generative grammar.* Cambridge: Cambridge University Press.

Tomasello, M. (2008). *The origins of human communication.* Cambridge, MA: MIT Press.

Trubetzkoy, N. (1958). *Grundzüge der phonologie.* Göttingen: Vandenhoeck & Ruprecht. (Originally published in 1939)

Tuckute, G., Kanwisher, N., & Fedorenko, E. (2024, August). Language in brains, minds, and machines. *Annual Review of Neuroscience, 47*(1), 277–301.

van Fraassen, B. C. (1980). *The scientific image.* Oxford: Oxford University Press.

van Rooij, I., & Baggio, G. (2020). Theory before the test: How to build high-verisimilitude explanatory theories in psychological science. *Perspectives on Psychological Science, 15*(4), 1044–1056.

Vargha-Khadem, F., Gadian, D. G., Copp, A., & Mishkin, M. (2005). FOXP2 and the neuroanatomy of speech and language. *Nature Reviews Neuroscience, 6*(2), 131–138.

Vaswani, A., Shazeer, N., Parmar, N. et al. (2023). *Attention is all you need.* https://arxiv.org/abs/1706.03762.

Veltman, F. (1996). Defaults in update semantics. *Journal of Philosophical Logic, 25*(3), 221–261.

von Humboldt, W. (1836). *On language: On the diversity of human language construction and its influence on the mental development of the human species* (M. Losonsky, Ed. & P. Heath, Trans.). Cambridge: Cambridge University Press.

Vong, W. K., Wang, W., Orhan, A. E., & Lake, B. M. (2024). Grounded language acquisition through the eyes and ears of a single child. *Science, 383*(6682), 504–511.

Weisberg, M. (2007). Three kinds of idealization. *The Journal of Philosophy, 104*(12), 639–659.

Weisberg, M. (2013). *Simulation and similarity: Using models to understand the world.* Oxford: Oxford University Press.

Wetzel, L. (2009). *Types and tokens: On abstract objects.* Cambridge, MA: MIT Press.

Wigner, E. P. (1960). The unreasonable effectiveness of mathematics in the natural sciences. *Communications on Pure and Applied Mathematics, 13*, 1–14.

Wilcox, E., Levy, R., Morita, T., & Futrell, R. (2018, January). What do RNN language models learn about filler–gap dependencies? In T. Linzen, G. Chrupała, & A. Alishahi (Eds.), *Proceedings of the 2018 EMNLP workshop BlackboxNLP: Analyzing and interpreting neural networks for NLP* (pp. 211–221). Brussels: Association for Computational Linguistics.

Worrall, J. (1989). Structural realism: The best of both worlds? *Dialectica, 43*(1–2), 99–124.

Yalcin, S. (2018). Semantics as model-based science. In D. Ball & B. Rabern (Eds.), *The science of meaning: Essays on the metatheory of natural language semantics* (pp. 334–360). Oxford: Oxford University Press.

Yan, K., & Hricko, J. (2017). Brain networks, structural realism, and local approaches to the scientific realism debate. *Studies in History and Philosophy of Science Part C: Studies in History and Philosophy of Biological and Biomedical Sciences, 64*, 1–10.

Yang, Y., & Piantadosi, S. T. (2022). One model for the learning of language. *Proceedings of the National Academy of Sciences, 119*(5), 1-12, e2021865119.

Yax, N., Oudeyer, P.- Y., & Palminteri, S. (2024). *PhyloLM : Inferring the phylogeny of large language models and predicting their performances in benchmarks.* https://arxiv.org/abs/2404.04671.

Yngve, V. H. (1996). *Linguistics as a science.* Indiana: Indiana University Press.

# Acknowledgements

I would like to thank Jacob Stegenga for his guidance and support during the process of writing this Element. I would also like to thank two excellent referees for their careful and considered engagement with the text and extremely useful suggestions for its improvement. Special thanks to Fintan Mallory for the great feedback!

# Cambridge Elements ≡

# Philosophy of Science

### Jacob Stegenga
*NTU Singapore*

Jacob Stegenga is a Professor at NTU Singapore, and previously taught at the University of Cambridge. He has published widely in philosophy of science and philosophy of medicine, and is the author of *Medical Nihilism*, described as 'a landmark work', *Care and Cure: An Introduction to Philosophy of Medicine*, and a book to be published in 2025 titled *Heart of Science*.

### About the Series

This series of Elements in Philosophy of Science provides an extensive overview of the themes, topics and debates which constitute the philosophy of science. Distinguished specialists provide an up-to-date summary of the results of current research on their topics, as well as offering their own take on those topics and drawing original conclusions.

**Cambridge Elements**

## Philosophy of Science

### Elements in the Series

*Modelling Scientific Communities*
Cailin O'Connor

*Logical Empiricism as Scientific Philosophy*
Alan W. Richardson

*Scientific Models and Decision Making*
Eric Winsberg and Stephanie Harvard

*Science and the Public*
Angela Potochnik

*Feminist Philosophy of Science*
Anke Bueter

*Abductive Reasoning in Science*
Finnur Dellsén

*Climate Science*
Wendy S. Parker

*The Social Dimensions of Scientific Knowledge: Consensus, Controversy, and Coproduction*
Boaz Miller

*Scientific Realism*
Timothy D. Lyons

*Science, Pseudoscience, and the Demarcation Problem*
Dániel Bárdos and Adam Tamas Tuboly

*Underdetermination and Theoretical Virtues*
Dana Tulodziecki

*The Philosophy of Linguistics*
Ryan M. Nefdt

A full series listing is available at: www.cambridge.org/EPSC

Printed by Integrated Books International,
United States of America